MATERIALS TO RESIST WEAR

A Guide to their Selection and Use

Tom Farris

THE PERGAMON MATERIALS ENGINEERING PRACTICE SERIES

OTHER TITLES IN THE SERIES

MATERIALS TO RESIST WEAR

A Guide to their Selection and Use

A. R. LANSDOWN

and

A. L. PRICE

Swansea Tribology Centre, U.K.

PERGAMON PRESS

OXFORD · NEW YORK · TORONTO · SYDNEY · FRANKFURT

U.K.	Pergamon Press Ltd., Headington Hill Hall, Oxford OX3 0BW, England
U.S.A.	Pergamon Press Inc., Maxwell House, Fairview Park, Elmsford, New York 10523, U.S.A.
CANADA	Pergamon Press Canada Ltd., Suite 104, 150 Consumers Road, Willowdale, Ontario M2J 1P9, Canada
AUSTRALIA	Pergamon Press (Aust.) Pty. Ltd., P.O. Box 544, Potts Point, N.S.W. 2011, Australia
FEDERAL REPUBLIC OF GERMANY	Pergamon Press GmbH, Hammerweg 6, D-6242 Kronberg, Federal Republic of Germany
JAPAN	Pergamon Press Ltd., 8th Floor, Matsuoka Central Building, 1-7-1 Nishishinjuku, Shinjuku-ku, Tokyo 160, Japan
BRAZIL	Pergamon Editora Ltda., Rua Eça de Queiros, 346, CEP 04011, São Paulo, Brazil
PEOPLE'S REPUBLIC OF CHINA	Pergamon Press, Qianmen Hotel, Beijing, People's Republic of China

First edition 1986

Library of Congress Cataloging-in-Publication Data

Lansdown, A. R.
Materials to resist wear
(The Pergamon materials engineering practice series)
Bibliography: p.
1. Mechanical wear. 2. Hard materials.
I. Price, A. L. II. Title. III. Series.
TA418.4.L36 1986 620.1'1292 85–19168

ISBN 0–08–033442–3
ISBN 0–08–033443–1 (pbk.)

British Library Cataloguing in Publication Data

Lansdown, A. R.
Materials to resist wear.—(The Pergamon materials engineering practice series)
1. Mechanical wear 2. Materials—Deterioration
3. Friction
I. Title II. Price, A. L.
620.1'122 TA418.72

ISBN 0–08–033442–3
ISBN 0–08–033443–1 (pbk.)

Printed in Great Britain by A. Wheaton & Co. Ltd., Exeter

Materials Engineering Practice

FOREWORD

The title of this series of books "Materials Engineering Practice" is well chosen since it brings to our attention that an era where science, technology and engineering condition our material standards of living, the effectiveness of practical skills in translating concepts and designs from the imagination or drawing board to commercial reality, is the ultimate test by which an industrial economy succeeds.

The economic wealth of this country is based principally upon the transformation and manipulation of *materials* through *engineering practice.* Every material, metals and their alloys and the vast range of ceramics and polymers has characteristics which require specialist knowledge to get the best out of them in practice, and this series is intended to offer a distillation of the best practices based on increasing understanding of the subtleties of material properties and behaviour and on improving experience internationally. Thus the series covers or will cover such diverse areas of practical interest as surface treatments, joining methods, process practices, inspection techniques and many other features concerned with materials engineering.

It is to be hoped that the reader will use this book as the base on which to develop his own excellence and perhaps his own practices as a result of his experience and that these personal developments will find their way into later editions for future readers. In past years it may well have been true that if a man made a better mousetrap the world would beat a path to his door. Today, however, to make a better mousetrap requires more direct communication between those who know how to make the better mousetrap and those who wish to know. Hopefully this series will make its contribution towards improving these exchanges.

MONTY FINNISTON

Preface

This book is intended to provide industrial engineers and designers with simple advice on the selection of wear-resistant materials. It is not intended to explain in detail the various mechanisms or classifications of wear, or the reasons for the behaviour of different materials under different wear situations. A brief review of wear is included but for a deeper understanding of wear the reader is referred to the many excellent textbooks which exist, and which are included in the bibliography.

The effects of lubricants on wear are complex. In order to keep the detailed guidance as simple as possible the effects of lubricants have been described separately and excluded from the detailed selection sections. Users who are concerned with selecting materials for a lubricated situation should therefore read Chapter 11 as well as the appropriate selection section.

Liquids may be present in wear situations without acting as lubricants. Examples are the transport of abrasive materials in the form of slurries through pipes, and the presence of environmental liquids such as rain-water, sea-water, fuels or process liquids, and the effects of these are also described.

Corrosive wear is a process in which some type of chemical reaction contributes strongly to the wear process. It can be very complex and relatively little work has been published about it. It is therefore not possible to give simple detailed guidance to the selection of materials to resist corrosive wear, but some general guidelines are given in Chapter 9.

The selection of a material to resist wear depends on the type of wear which is taking place. It is not always enough to use the hardest material which is available. It follows that it is necessary to recognise the type of wear before a material selection can be made.

Unfortunately none of the existing classifications of wear type is helpful to the non-specialist user. "Mild wear" is not necessarily slower than "severe wear". "Scuffing" and "scoring" mean the same thing in the United States and different things in the United Kingdom. "Abrasion" means different things to different specialists. Furthermore, even for those who understand clearly what is meant by these terms, it is by no means easy to recognise which type of wear is taking place in a particular case.

The approach which is taken in this guide is therefore to classify the wear situation rather than the wear type. Each wear situation is defined in Chapters 1 and 2 in simple terms which will be easily recognised by every user. For each category of wear situation there then follows a separate section (Chapters 3–9) giving detailed guidance to the selection of resistant materials.

In addition to lubrication and corrosive wear there are several topics which were considered to merit separate description. These include wear testing, wear-resistant coatings, and some sources of the materials recommended throughout the guide. Finally there is a bibliography of useful publications.

Thanks are due to the Department of Industry which made available a grant which enabled the original work on this project to be undertaken, to Mr. P. L. Hurricks who contributed to the early work on the project, to Dr. T. S. Eyre who supplied some of the illustrations, and to Professor F. T. Barwell, who read the draft and provided many useful suggestions.

Contents

Chapter

General Review of Wear

Chapter 1

General Review of Wear

1.1 INTRODUCTION

Wear can be defined as the progressive loss of material from the operating surface of a body occurring as a result of relative motion at the surface. The problem of wear arises wherever there are load and motion between surfaces, and is therefore important in engineering practice, often being the major factor governing the life and performance of machine components.

As wear occurs in a wide variety of industries, it is also an item causing major annual expense throughout industry. The factors which affect wear are numerous, such as the type and method of loading, speed, temperature, materials, whether there is lubricant present and, if so, what type and in what quantity, and the chemical nature of the environment. The variety of conditions, therefore, makes the study of wear a very complex subject.

In May 1964 an active co-operative programme on the comparison of sliding wear test results was initiated under the auspices of the Organisation for Economic Co-operation and Development (O.E.C.D.). The purpose of the programme was essentially to establish a basis for reporting and comparing results obtained by various test methods; in particular, it tried to establish whether or not a great number of participants could agree on the wear behaviour of a number of different material combinations, using different test rigs.

The overall result of the programme was disappointing in that the agreement between the results obtained in different laboratories was extremely poor. However, it became apparent that many of the participating laboratories agreed on the types of wear mechanisms that were operative and the reproducibility of the results was reasonable if only one mechanism was taking place. Seemingly minor factors, among which the energy balance at the friction interface is probably the most important, may have a considerable influence on the type of wear mechanism that predominates, or on the relative importance of the different mechanisms that may be operative successively or simultaneously. Because of the powerful influence of minor factors on the type of wear mechanism which predominates, a very large scatter is bound to occur between experiments with specimens of varied geometry. Other factors brought to light by the programme were that laboratory equipment of a general type should only be used to investigate the influence of different parameters on the transition from the mild or no-wear region to the severe wear region, rather than for measuring absolute values of friction and wear; and test rigs should be so constructed that the conditions found in service are simulated as closely as possible.

The résumé of the O.E.C.D. programme indicated that there was little success in correlating different types of wear test, and only served to produce a better understanding of the difficulties that exist when a variety of wear mechanisms are operating simultaneously.

Since then a great deal of fundamental research has been carried out on wear, especially in the understanding of individual mechanisms and the effect of interaction between them, but the O.E.C.D. findings are still valid.

1.2 NATURE OF SURFACE CONTACT

Solid surfaces in engineering are never smooth on a microscopic scale, but are covered with asperities, and these are usually between 0.1 μm (4 μin) and a few microns high.

When two surfaces are loaded together they make contact initially at the tips of a few asperities. As the load increases, the contacting asperities deform so that the contact area on each increases, and at the same time additional asperities make contact. The total actual area of contact is equal to the average area of an asperity contact multiplied by the number of asperity contacts.

In practice this actual or "real" area of contact will rarely increase to such an extent that the whole of the two surfaces are in contact, and so it can be assumed that any two surfaces loaded against each other are only in actual contact at a greater or lesser number of asperities.

1.3 THE NATURE OF WEAR

The subject of wear is complicated by a confusion of nomenclature and the lack of good definition of the different types of wear found in engineering situations. The various modes of wear include adhesion (including scuffing, galling, welding, scoring, wiping and smearing), abrasion, fatigue, erosion and corrosion. Corrosion is a complex phenomenon in its own right, strongly influenced by such factors as environment, materials and the presence or absence of a film such as a lubricant, or other contaminants. Corrosion is not a primary subject of the survey, but a review of corrosive wear is given in Chapter 9.

1.4 ADHESIVE WEAR

When two surfaces are loaded against each other the whole of the contact load is carried on the very small area of the asperity contacts. The real contact pressure at these asperities is very high, and adhesion takes place between them. Ductile materials, such as metals and plastics, may also exhibit some diffusion or crystal growth at these junctions, thus causing stronger adhesion.

If one of the surfaces is moved sideways over the other, the adhesive junctions will break. As sliding continues, fresh junctions will form and be ruptured in turn.

If the adhesive strength of the junction is less than the cohesive strengths of the materials forming the asperities, then the junction will rupture at the original point of contact, and there will be no loss of material from either of the two surfaces. If, on the other hand, the adhesive strength is greater than the cohesive strength of either of the two materials, then the junction will rupture within the weaker asperity. The two possibilities are shown in Fig. 1.1, where Path 1 represents rupture along the original contact and Path 2 represents rupture within the weaker asperity.

It would seem reasonable to expect that rupture would normally take place at the original contact, since this is the shortest path and is weakened by contaminants and mismatching of the crystal or grain structures. In practice, some direct investigations showed that a typical junction break took place very close to, but not at, the original contact surface, so that small numbers of atoms were transferred from one surface to the other.

Such a process represents wear of the surface from which the atoms have been transferred, although it would be insignificant from the

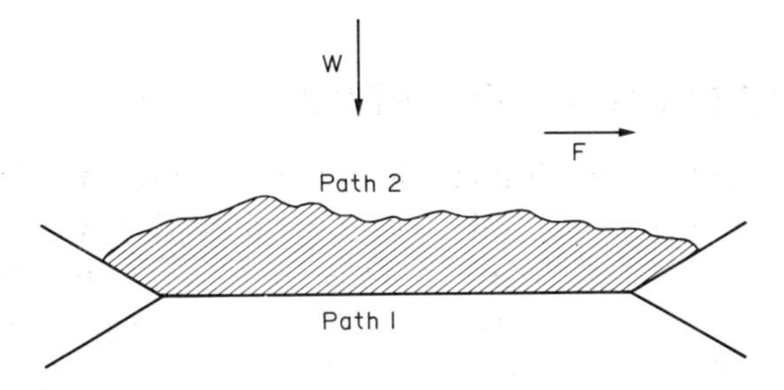

FIG. 1.1 ALTERNATIVE RUPTURE PATHS FOR AN ASPERITY JUNCTION

engineering aspect. However, once transfer has taken place to an asperity on the opposite surface, the next contact will be between similar materials on both surfaces, and will usually give much stronger adhesion. This will, in turn, tend to give a greater amount of transfer.

Although this transfer represents a form of wear, more typical wear results in the formation of separate particles, or wear debris. The process by which debris is formed in adhesive wear probably involves a weakening of an asperity tip due to repeated compression and tension. Eventually the impact against an opposing asperity is sufficient to break the weakened particle away from the surface to form a wear particle. In other words, a fatigue process is involved.

The rate of wear debris formation and the size of the wear particles depend on the severity of the adhesion which takes place between the asperities.

1.4.1 Types of adhesive wear pattern

When two solid surfaces are rubbed together the wear can progress in different ways. At a constant load and speed the wear data can be represented graphically as the wear volume versus sliding distance, the wear rate being the slope of the line in each case.

The wear behaviour may be either a simple direct proportionality between wear volume and sliding distance as in Fig. 1.2, or as in Fig. 1.3 an initial high wear rate preceding a change to a lower constant rate. The initial high wear rate is usually referred to as "running-in" or transient wear and the final constant rate as equilibrium wear.

Plotting equilibrium wear rate against load for constant speed, a graph as in Fig. 1.4 is obtained. Similar nonlinear situations arise when load is constant and speed is varied, as in Fig. 1.5. The transitions T_1 and T_2 separate the wear pattern into distinct regions, termed mild wear and severe wear, and in every material (metallic, polymeric or ceramic) studied over a wide range of load and speed, at least one such transition has been detected.

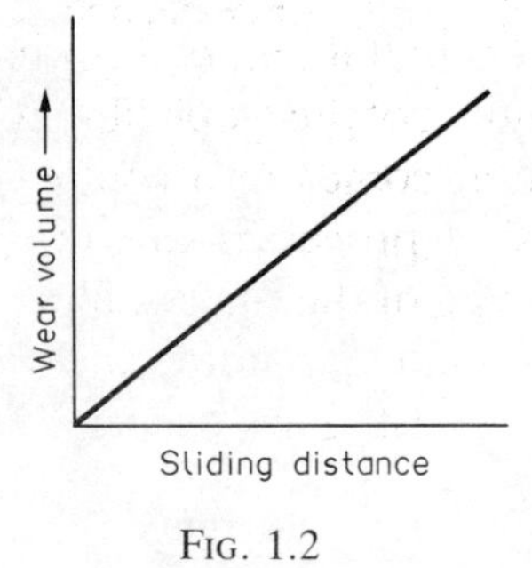

FIG. 1.2

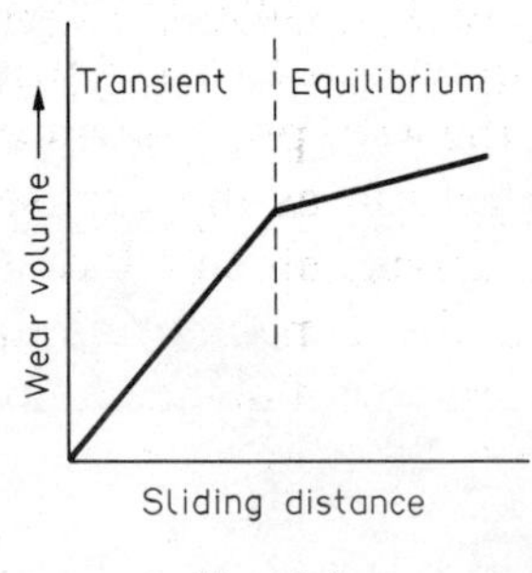

FIG. 1.3

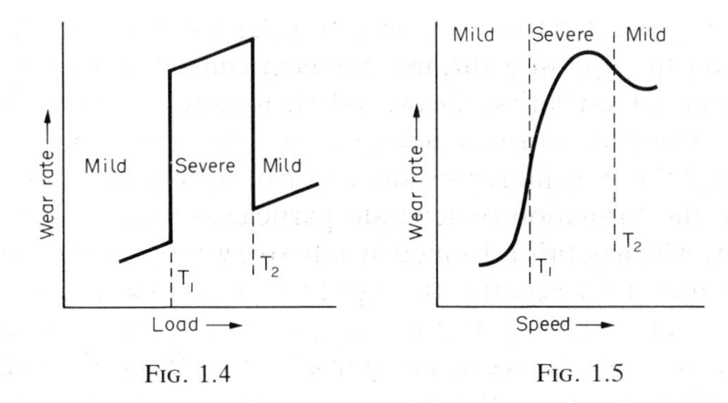

FIG. 1.4

FIG. 1.5

1.4.2 Mild and severe wear

For many metals, the type of wear, i.e. whether mild or severe, depends upon the prevailing rubbing conditions. In mild wear the surfaces are often visibly oxidised. It is these surface oxide films which give protection from serious damage because the adhesion between oxide layers at the asperity tips is weak and shears easily. Sliding conditions which favour oxide film formation during the rubbing process, i.e. low loads and speeds, are thus more likely to support mild wear. The wear detritus in mild wear generally manifests itself as a fine fully-oxidised powder which is obviously non-metallic.

If the sliding velocity, temperature or load is increased, a situation may be reached where the oxide films cannot form properly between successive contacts, and the wear process can change to severe wear. In severe wear the debris is almost entirely metallic, consisting of large metallic fragments which have been torn away due to the strong adhesive junctions which form with an appreciable degree of intermetallic contact in the absence of adequate oxide films, thus giving a rough and torn appearance to the surfaces.

In continuous rubbing the factors which determine whether a system is in a suitable or safe operating range are therefore essentially dynamic. Most metals are covered with a surface layer, usually oxide, which protects the surfaces from direct metal to metal contact. With rubbing, the wear process causes removal of the oxide film, which will begin to reform as the exposed metal comes into contact with an oxidising environment. Two opposing processes are therefore in operation, the damaging and the healing of the surface film.

If the process of damaging predominates, then the wear becomes worse and severe wear takes place, whilst if the process of healing predominates, the wear rate is kept at a low level and mild wear occurs.

The rate of damage will increase with increase in load and sliding speed, whilst the rate of healing will increase with temperature. Increasing the sliding speed, however, will also increase the frictional heating effect, with the result that the damage caused by increased speed may be overtaken by the healing effect associated with temperature and the system reverts to mild wear. Severe wear is therefore often found between two regions of mild wear, as shown in Figs. 1.4 and 1.5.

1.4.3 The influence of surface films

Both the friction between rubbing surfaces and the wear which occurs can be reduced by the use of low shear-strength surface films on a hard substrate. This technique allows the applied load to be supported through the soft film by the base material, while any shear takes place within the surface film. Many types of film are suitable, including oxides, chlorides, sulphides, and other reaction products, and also certain soft metals such as lead and silver, and many non-metallic materials.

Oxides are the most common protective films, and they form naturally in most engineering environments. To be effective as a protection against wear, an oxide film must have certain properties, these being:

1. Good adherence.
2. Limited thickness.
3. Low shear strength.
4. Compressive strength less than or similar to that of the substrate metal.

Good adherence is obtained when oxide–metal specific volume ratios are greater than one. If the ratio is less, then the oxide films tend to flake and crack off. Metals with favourable volume ratios are copper, beryllium, zinc aluminium, silicon, lead, chromium, molybdenum, iron, cobalt and nickel. Titanium has an unfavourable ratio.

The measure of protection offered will largely depend upon the relative mechanical properties of the metal and its oxide, similar properties being necessary in order to prevent any break up of oxide under load (both metal and oxide are able to deform together). The pure metal and oxide properties of some common engineering metals are shown in Table 1.1; it can be seen that the greater the difference in hardness between metal and oxide, the lower is the load required to produce appreciable metal contact.

TABLE 1.1 COEFFICIENT OF FRICTION OF CLEAN AND OXIDE-COATED METALS AT 25°C

Metal	Hardness, kg/mm^2		Load at which appreciable metal contact occurs, g	Coefficient of friction	
	Metal	Oxide		Metal on metal	Oxide on oxide
Gold	20	—	0	2	—
Silver	26	—	0.003	1	0.8
Tin	5	1650	0.02	1	1
Aluminium	15	1800	0.2	1.2	0.8
Zinc	35	200	0.5	0.8	1.2
Copper	40	130	1.0	1.6	0.8
Iron	120	150	10.0	0.6	1.0
Chromium	800	—	1000.0	—	0.4

The formation of oxides at low temperatures has usually been found to be more effective than at high temperatures. However, high frictional temperatures during dry rubbing may lead to a complete change of crystal phase and intense hardening of surface layers which can cause change in wear behaviour.

1.4.4 Influence of the chemical constitution of the sliding pair

The ability or inability of a metallic pair to produce bonding is important in dry or poorly-lubricated sliding applications. Pairs of metals which produce the fewest and weakest bonds would be of benefit in sliding applications. The fewer the bonds, the fewer the number of possible points available for welding; the weaker the bonds the smaller the tendency for metal transfer and work-hardening of the asperities. To satisfy these conditions it has been found that the sliding pair should:

(a) Be mutually insoluble.
(b) Have at least one of the metals from the B sub-group of the periodic table (see Table 1.2).

The mutual solubility will determine the number of junctions likely to form and the bonding characteristics will determine their strength.

An indication of the influence of these two conditions on sliding metal pairs is given in Table 1.2, where a number of different metal sliders were loaded against a rotating steel disc, the scuff resistance being measured as a function of the load-carrying capacity.

TABLE 1.2 SCUFFING RESISTANCE OF ELEMENTS AGAINST 1045-STEEL

Good	Fair	Poor	Very poor
Germanium*	Carbon	Magnesium	Beryllium
Silver*	Copper* ←	Aluminium	Silicon
Cadmium* ←	Selenium*	→ Copper*	Calcium
Indium*	→ Cadmium*	Zinc*	Titanium
Tin*	Tellurium*	Barium	Chromium
Antimony*		Tungsten	Iron
Thallium*			Cobalt
Lead*			Nickel
Bismuth*			Zirconium
			Columbium
			Molybdenum
			Rhodium
			Palladium
			Cerium
			Tantalum
			Iridium
			Platinum
			Gold*
			Thorium
			Uranium

* B-subgroup.

Metals which have the best scuffing resistance against steel are the B-subgroup metals which are either insoluble with iron or else form intermetallic compounds with iron. The metals showing a poor or very poor scuffing resistance were found to be either soluble in iron or of a group other than a B-subgroup (copper being an exception by exhibiting some diversity in the results obtained).

1.4.5 Influence of the crystal structure

The type of crystal structure (i.e. whether hexagonal or cubic) inherent to the sliding metals greatly influences the friction and wear of the couple. The force required for the shearing of welds depends upon the plane along which shearing takes place. During sliding, shear forces in cubic crystals are normally greater than the corresponding shear forces in hexagonal crystals for two reasons. Firstly, cubic crystals work-harden at a greater rate during sliding, and secondly there are planes of "easy slip" in hexagonal crystals. During sliding it has been found that hexagonal metals undergo recrystallisation and preferred orientation for easy slip at the surface. For sliding applications metals which remain in a hexagonal crystal form over the entire operating temperature range are therefore beneficial in reducing surface welding. Maintenance of the hexagonal crystal form can be brought about by alloying additions.

Grain size is also important. The grains of a polycrystalline material are influenced by their neighbours during deformation, unlike single crystals which have free boundaries. For polycrystalline materials the constraining action by neighbouring grains is least when the average grain diameter is much greater than the microscopic areas of contact. Therefore contact over a large number of grains will greatly reduce the rate at which wear occurs, and a large grain size is undesirable.

1.5 ABRASIVE WEAR

An alternative effect is possible when two asperities interact. Instead of both asperities deforming, and probably adhering, it is possible for a harder asperity to plough or cut through a softer one. It is also possible that if two hard or brittle asperities of similar hardness interact, one of them may be ploughed aside or broken off if, for some reason, it experiences a higher stress than the other.

The result of cutting or ploughing wear is a significant loss of material from one of the surfaces, which develops a scored, grooved, or machined appearance. The appearance of the wear debris depends on the material being worn and the temperature. If the debris arises at fairly low temperatures from steel or bronze it will generally have the appearance of machining swarf, and the size will depend on the size of the cutting asperities. If the debris arises from a soft surface, especially at elevated temperatures, it may develop a partly melted appearance, or may re-attach to one of the surfaces.

A general name for the wear produced by cutting or ploughing is "abrasive wear", which can be defined as wear due to the penetration and ploughing-out of material from a surface by another body. Two general situations which exist for this type of wear are two-body abrasive wear and three-body abrasive wear. If the abrading contact is a free grit from an external source or a wear particle which has been generated within the system and which is loaded between two surfaces, then the situation is refered to as three-body abrasion. If the abrading contact is a surface asperity on the mating part, the situation is referred to as two-body abrasion, or cutting wear.

For a material to be scratched or worn, the abrasive must normally be harder than that material, although some wear of the harder material does occur even when relatively soft minerals rub against hard materials. There is, however, a marked drop in wear rate when the hardness of the abrasive is less than that of the material it abrades.

In the two-body abrasive condition, the damage caused by abrasion can, therefore, be controlled to some extent by the relative hardness and surface finish of the mating pair, although other factors such as elastic modulus, micro-structure, etc. have some influence. It is common practice to make one surface considerably softer than the other in order to confine any damage to the component which is easily renewable. The harder of the two surfaces is also given a good smooth finish in order to minimise abrasion by protruding asperities.

One special case of abrasive wear can arise in which the harder of the two surfaces is abraded. In the course of formation a wear particle can be work-hardened so that it is harder than the surface from which it was formed. If such a particle becomes partly embedded in the softer counterface, it can form a cutting tip and cause severe cutting wear of the harder surface. A particularly severe form of cutting wear is known as "wire-wooling", and has, in the past, caused serious damage and failures of high-speed turbines.

Three-body abrasive situations are common in industrial applications due mainly to the environmental conditions which exist in industrial plants, the major part of the wear-producing abrasives coming from the surrounding atmosphere in the form of dust or grit. It must be noted, however, that the external environment is not the only source of such hard particles and that other sources such as internally generated wear debris or corrosive wear products may occur.

Specific properties associated with the abrasive particles have significant effects on surface wear. Coarse particles produce rougher surfaces and generally more wear than fine ones. Hardness is important in that a material must be harder than a mating surface in order to cause significant abrasion. Particle hardness relative to the two surfaces will also determine whether the debris remains free or whether it becomes embedded in the softer surface to act as a lap. It has also been found that particle shape exerts some influence in that angular particles produce greater wear than rounded ones.

1.6 SURFACE FATIGUE

A further mechanism of wear is that of surface fatigue, which results when repeated sliding, rolling, or impacting motions subject a surface to repeated cycles of stress. The stress cycles initiate cracks in or near the surface which, with time, spread, link up and form discrete particles, free to move between the contacting surfaces. During the operation of the wear mechanism, continual stressing and unstressing of the material takes place and for some time the material appears to be quite unaffected by this stress cycling. Eventually, perhaps after many hours of operation, a particle will detach from the surface. This particle can contribute to three-body abrasion, and rapid deterioration of the surface may follow, due to the flaking off of further fragments.

Many of the characteristics of surface fatigue wear are quite similar to those exhibited in the fatigue testing of bulk specimens, but there are two important differences. The first is that variations in the life are much greater in surface fatigue compared to the bulk case, e.g. in the rolling contact situation the ratio may be about 10 to 1. The second is concerned with the phenomenon of the fatigue limit stress observed in ordinary bulk testing, below which the material has an infinite fatigue life. In the case of surface fatigue no such limit apparently exists. These differences therefore make the life prediction of an individual system under cyclic stress loading difficult, particularly with respect to design.

This phenomenon is not the same as the fatigue mechanism which contributes to adhesive wear. It can arise in lubricated systems where there is no asperity contact, and therefore no adhesion.

Surface fatigue is a particularly important wear mechanism in ball and roller bearings, where it is usually called "rolling-contact fatigue".

1.7 EROSION

Erosion is the term applied to the damage produced by the impingement of sharp particles on an object and is closely analogous to abrasion. The main difference between the two wear types is that with erosion the surface roughness produced may be relatively more severe than with abrasion, due to removal of material from low points on the surface by the striking particles. Erosion, although usually considered undesirable, has useful applications in such processes as sand blasting, abrasive deburring and erosive drilling of hard materials.

The understanding of the erosion phenomenon may be divided into two fairly distinct areas. The first is concerned with the determination from the fluid flow conditions of the properties of the particles striking the surface, e.g. number, direction and velocity, and this is a problem of fluid mechanics. The second involves the amount of material removed from the surface being attacked by the particles. Removal of material occurs by two wear mechanisms, one being the repeated deformation during collisions which results in the breaking loose of pieces of material, the other being the cutting action of free-moving particles. In practice, these two wear mechanisms occur simultaneously. Material removal is thus governed to a large extent by the type of material undergoing the erosive attack. Ductile materials undergo weight loss by a process of plastic deformation, the material being removed by the displacing or cutting action of the eroding particles. Brittle material, however, is removed by the intersection of cracks which radiate out from the points of impact of the eroding particles. There are obviously materials which fall between these two categories and therefore material removal would involve some combination of these two wear processes.

1.8 PRINCIPAL FACTORS INFLUENCING WEAR

Apart from the effect of lubrication, which is described in Chapter 11, there are several major factors which influence wear.

Hardness. In general, increasing hardness decreases the wear of a material, but there is no simple relationship between the two. In adhesive wear, hardnesses above 700 VPN tend to suppress severe wear, so that the wear is restricted to the mild form. In abrasive wear, there is some evidence that the wear rate of commercially pure metals and heat-treated steels is inversely proportional to their hardness. With surface fatigue and erosion the effect of hardness is far less straightforward.

Load. There is also a general tendency for wear rate to increase as the load increases, although as shown in Fig. 1.4 there can be stepwise transitions in adhesive wear which reverse the trend. There is also a critical point in most systems beyond which an increase in load leads to seizure rather than to an increase in wear rate.

Speed. Wear rate can change considerably with change in speed, but there is no general relationship between speed and wear rate. An increase in speed can lead to an increase or a decrease in wear, depending on the effect on the temperatures of the surfaces.

Surface roughness. In two-body abrasive wear, the roughness of the harder surface has a major influence on the wear rate. In adhesive wear, initial roughness may lead to severe damage or may gradually be reduced to an equilibrium "run-in" level. There is also evidence that surfaces which are initially very smooth can become roughened in service until they reach a similar equilibrium "run-in" roughness.

Temperature. In general an increase in temperature tends to produce an increase in wear rate, because with increasing temperature the materials involved become softer. In adhesive wear the types of transition shown in Figs. 1.4 and 1.5 can occur, so that an increase in temperature may sometimes cause a decrease in wear. In the severe form of adhesive wear called scuffing, there is strong evidence that the onset of scuffing is determined directly by increase in temperature.

Chapter 2

Diagnosing the Cause and Nature of Wear

2.1 IDENTIFYING THE CAUSE OF WEAR

It is important for both the designer and the user of equipment to establish the nature of the wear taking place in a given situation, because the action taken, and the materials selected to reduce wear, will depend critically on the nature of the wear process.

For the user, however, it may be important to consider the cause of wear in a much more general sense, because the most common causes of excessive wear are not related to the selection of the best wear-resistant materials. They include supply of incorrect material, wrong heat treatment, faulty manufacture, and changes in the operating environment.

The key in Fig. 2.1 shows how the most general preliminary assessment of the cause of excessive wear should be made. It can be stated quite categorically that if a system has a history of satisfactory wear performance and it begins to give excessive wear, then some change has taken place which is at least a partial cause of the deterioration. It follows that if the change can be identified this will be an important step in solving the wear problem.

It is also true that if a standard piece of equipment is showing excessive wear in one application but not in others, then there is some difference between the applications which will account for the difference in performance. In this case, however, it will often be more difficult to establish the difference in operating conditions, because the various applications may be under the control of different managements.

In practice a serious increase in wear can be caused by a difference in an operating condition which is slight or even undetectable. Nevertheless it is important to carry out this analysis, because it will often lead to the simplest and quickest solution of the wear problem.

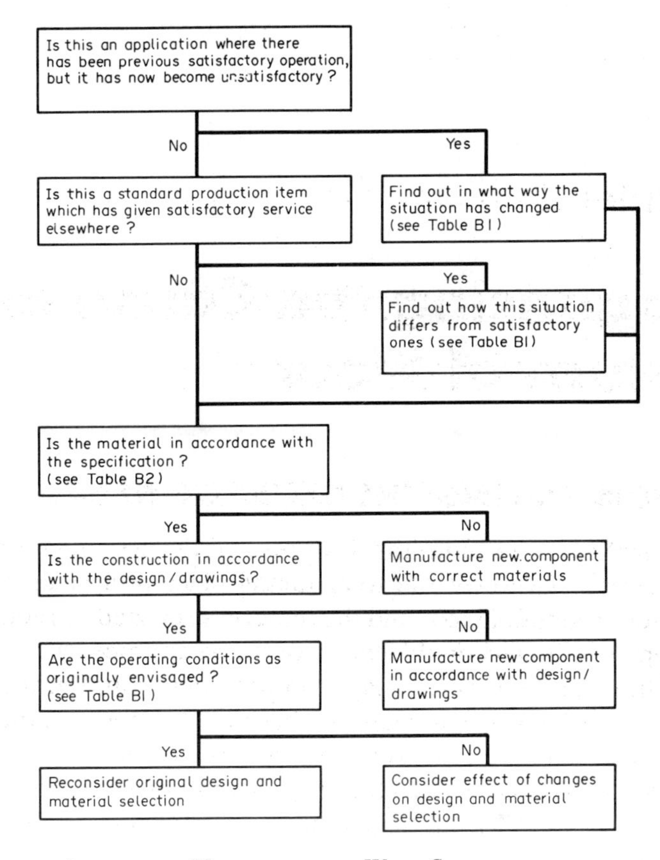

FIG. 2.1 KEY TO ANALYSIS OF UNSATISFACTORY WEAR SITUATION

Table 2.1 lists the operating conditions which are important in controlling the amount and nature of wear.

When standard equipment is being used, a very common cause of exceptional wear is that there is a fault in the materials used. Table 2.2 lists some of the common material faults which arise, and the techniques available for assessing them.

Material problems often arise when there has been a change of material supplier, and these may not always be related to the supply of material which is off specification. For example a change of supplier may lead to a change in concentration of an element such as vanadium which is only specified to a low maximum concentration but whose presence in trace amounts can significantly alter hardening characteristics.

If neither the analysis of operating conditions nor material faults leads to a solution to the wear problem, it becomes necessary to reconsider the design and the selection of materials.

TABLE 2.1 CONDITIONS AFFECTING NATURE AND EXTENT OF WEAR

1. Load
2. Speed
3. Vibration or dynamic loading
4. Temperature
5. Presence of loose abrasives
6. Nature of loose abrasives
7. Nature of gaseous environment
8. Contaminants
9. Lubrication
10. Damage in manufacture or assembly

TABLE 2.2 COMMON MATERIAL FAULTS AFFECTING WEAR BEHAVIOUR

Fault	Analytical techniques
1. Wrong alloy composition	Spectrographic analysis, atomic absorption spectrometry, wet chemical analysis, metallography, electron probe micro-analysis, Auger spectrometry, energy dispersive X-ray analysis.
2. Wrong heat treatment or micro-structure	Metallography, electron microscopy, hardness, micro-hardness, impact testing, tensile testing.
3. Poor quality	Metallography (for inclusions, cracks, corrosion, etc.), crack detection.
4. Surface defects	Wrong surface finish, wrong orientation with respect to grinding or turning marks, rolled-in inclusions, etc.
5. Residual stress	Can only be assessed by destructive techiques.

2.2 RECOGNISING THE WEAR SITUATION

In order to make the best selection of material to resist wear, it is necessary to identify the type of wear which is taking place.

Wear has been classified in many ways, and over thirty different titles have been used for different types of wear. Some of these titles, such as adhesive wear and fatigue, relate to the mechanism by which wear occurs. Such titles are very useful to the wear specialist, but are of limited use for the non-specialist user.

Other categories, such as sliding wear and impact wear, are related to the situation in which the wear is taking place. These terms are often less precisely defined than the mechanistic terms, but are much easier for the non-specialist to recognise. Because this book is intended for the non-specialist, a classification based on wear situations, instead of mechanisms, is used.

For the purpose of selecting wear-resistant materials, the various types of wear are divided in this book into seven categories. These are:

Sliding wear.
Fretting.
Three-body abrasion.
Gouging wear.
Low-stress abrasion.
Erosion.
Corrosive wear.

Each of these categories meets two requirements.

(i) It is based on an easily-recognised wear situation.
(ii) Clear guidance can be given to the selection of wear-resistant materials within the category.

Some rubbing situations may involve more than one of the categories. For example, a surface may suffer fretting caused by vibration while it is nominally stationary, and suffer abrasive wear while it is moving. In such cases the first priority will be to reduce the most damaging type of wear.

In the following pages the full titles and descriptions of the seven wear categories are given, and should enable each type to be recognised.

CATEGORY 1. *Sliding wear: two solid surfaces rubbing against each other with no abrasive particles present*

If it is known that two solid surfaces are rubbing together but it is not clear whether abrasive particles are having any important effect, examine the two surfaces.

If at least one of the surfaces has a welded, or torn appearance, there is severe adhesive wear which is in Category 1A (see Figs. 3.1 and 3.2).

If at least one of the surfaces has a melted, or wiped, or smeared appearance, the situation probably falls in Category 1A (see Fig. 3.4).

If one surface is hard and rough, and the other is softer and has scratches or scores corresponding with the roughness on the hard surface, this may be cutting or machining wear or two-body abrasion and comes in Category 1B. Alternatively, abrasive particles may become embedded in the softer surface and scratch or score the harder surface, and this will again be in Category 1B. (See Fig. 3.4 for examples of cutting or machining wear and two-body abrasion.)

This situation may change into three-body wear (Category 3) if hard particles of wear debris are loosely trapped between the surfaces and then themselves cause abrasion of one or both of the surfaces.

CATEGORY 2. *Fretting: wear associated with vibration or low-amplitude oscillation*

Fretting is a common wear phenomenon and can be very serious and expensive. Where vibration or small oscillations are known to be present the problem can be easily recognised, but it often arises where vibration is not obvious. For example when machinery is transported by road, rail or air, fretting can occur between any pair of stationary surfaces in contact, such as bearings, gears, or even structural joints. Fretting can sometimes occur in a machine where vibration is transferred from a distant source through floors or mountings.

The damaged surface shows a general pitted appearance which can sometimes be confused with corrosion. Where ferrous metals are involved, the presence of brown ferric oxide powder ("Cocoa") is very characteristic. (See Fig. 4.1 for examples of fretting.)

CATEGORY 3. *Three-body abrasion: two solid surfaces rubbing against each other with abrasive particles present*

In this situation, if both surfaces are hard they will show scratch or score marks corresponding with the sharp corners of the abrasive particles. Examples are guides for hydraulic rams in a foundry or steel mill, balls in a ball mill.

If one surface is soft it will show deep score marks which do not usually correspond with projections on the other surface, and there may be abrasive particles embedded in the surface. (See Fig. 5.1 for example of three-body abrasion.)

CATEGORY 4. *Gouging wear: abrasive contacting or rubbing on a surface under high loads*

This is the situation which arises in the blades of earth-moving equipment, where gravel or flints embedded in the ground strike a digger blade or ploughshare with high contact stresses, or on sinter decks in a steelworks, where the sinter falls heavily on the deck surface.

The worn surface shows considerable uneven loss of material, concentrated at the places where the greatest contact occurs, and there are deep score-marks, gouges or grooves on the surface of the worn areas. (See Fig. 6.1 for examples of gouging wear.)

CATEGORY 5. *Low-stress abrasion: abrasives sliding along a surface under fairly light loads*

This situation arises in chutes or conveyors carrying coke, sand, gravel, metal ores, etc., where each individual contact is mild and produces light scratching, but the very large number of separate contacts results in considerable loss of material.

The worn surface shows fairly uniform loss of material, and has a polished appearance with superimposed scratches. (See Fig. 7.1 for examples of low-stress abrasion.)

CATEGORY 6. *Erosion: abrasive particles striking a surface at an angle with relatively light loads*

Erosion occurs in pipes carrying slurries, in shot-blasting cabinets, in cyclones and hydrocyclones and in structures exposed to blowing sand or dust. (It can also be caused by drops of liquid carried in a gas, by high-velocity liquid streams, or by cavitation in a liquid.)

The damaged surface shows non-uniform loss of material, which may follow flow lines. The damage will sometimes increase as wear occurs, so that a plate may be worn through at one point and relatively unaffected elsewhere. The damaged surface is generally smooth and polished, but if the abrasive strikes at a high angle and is harder than the surface, there may be superimposed pits or embedded particles. (See Fig. 8.1 for examples of erosion.)

CATEGORY 7. *Corrosive wear: wear which is increased by the occurrence of chemical reactions*

The chemical reactions which contribute to corrosive wear are generally similar to those which would take place with the same materials in the same environment, but often the corrosion is greater when wear occurs and the wear is greater when corrosion occurs.

It follows that corrosive wear should often be suspected where the environment is known to be chemically reactive. It may not be easy to

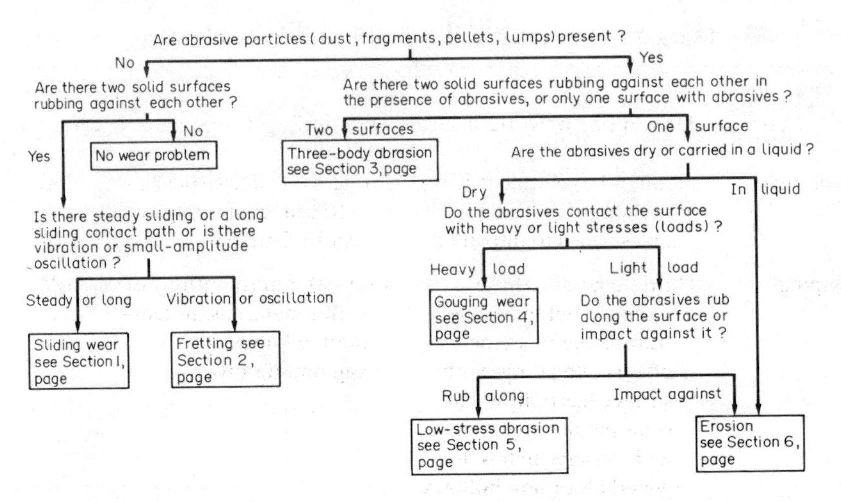

FIG. 2.2 SELECTION KEY

recognise because the surfaces may be bright and not obviously corroded. The best identification feature is often the highly corroded nature of the wear debris.

Remember that these categories are based on the situation in which wear is taking place. In order to decide the category into which a particular wear problem falls, it is only necessary to analyse the situation, and not the worn surface or the wear debris. This approach is made clearer in the Selection Key, Fig. 2.2.

However, a great deal of useful information can be obtained from an examination of the worn surfaces and the wear debris, and such examinations should be carried out wherever possible. In recent years several techniques have been developed, and some of these are described in Chapter 12.

Table 2.3 describes the appearance of the worn surfaces and the wear debris produced in the seven wear categories.

In summary, this chapter contains four ways to identify the category in which a particular wear problem belongs:

(i) The titles of the categories, which are descriptive of the wear situation.
(ii) The expanded descriptions of the categories.
(iii) The Selection Key (Fig. 2.2).
(iv) The descriptions of worn surfaces and wear debris (Table 2.3).

In addition, photographs of typical examples of worn surfaces are included in Chapters 3–9, which provide the detailed guidance to the selection of wear-resistant materials in each of the seven categories.

TABLE 2.3 DESCRIPTIONS OF WORN SURFACES AND DEBRIS

Type of Wear	Description of worn surfaces	Description of wear debris	Wear category
Adhesive	Small or large-scale tearing, transfer of material from one surface to the other.	Large irregular particles, $>10\ \mu m$, containing unoxidised metal.	1A
Wiping	Hard surface has little damage, but may show transferred metal or effects of heating. Soft surface has polished or semi-melted appearance, with surface material wiped off or into hollows.	Consists almost entirely of softer materials in semi-melted lumps or agglomerations.	1A
"Mild wear"	Fairly uniform loss of material, slight surface roughening.	Fine particles, $<1\ \mu m$, fully oxidised.	1A
"Two-body abrasion"			
A	Harder surface little or no damage. Softer surface has scores, grooves or scratches corresponding with rough asperities on harder surface.	Consists almost entirely of softer material, shaped like fine swarf, containing unoxidised metal.	1B
B	Softer surface contains embedded hard wear particles, may also be scored or wiped. Harder surface scored or grooved, may appear to be machined, with grooves corresponding with hard embedded particles in counterface.	Will contain material from harder surface in form of swarf, containing unoxidised metal. May also contain softer material in lumps.	1B
Fretting	Surfaces heavily pitted, pits may be fine or larger, or overlapping to produce rough areas. Surface of pits rough on microscopic scale, with oxidised appearance.	Fine, fully oxidised, if from ferrous metal will contain ferric oxide, Fe_2O_3, with appearance of cocoa. Sometimes spherical particles.	2

Continued overleaf

TABLE 2.3—*Continued*

Type of Wear	Description of worn surfaces	Description of wear debris	Wear category
Three-body abrasion	Surfaces have deep scratches or grooves, deeper on softer surface, similar on both surfaces if they are equally hard.	Fine, may contain some unoxidised metal, but usually lost amongst loose abrasive material.	3
Gouging wear	The one surface has scratches, gouges, and/or pits, depth depending on hardness of surface and nature of loose abrasives.	Not usually detectable amongst loose abrasive material.	4
Low-stress abrasion	The one surface will be unevenly worn, the worn areas having a polished appearance with fine superimposed scratches.	Usually difficult to detect amongst loose abrasive material.	5
Erosion	Surface very unevenly worn, worn area with smooth matt appearance, occasional embedded fine particles.	Usually difficult to detect amongst loose abrasive material.	6
Corrosive wear	Often, but not necessarily, showing signs of oxidation, corrosion or generally chemical reactions.	Always highly reacted; in other words no free metal present. Usually fine and amorphous.	7

Chapter 3

Sliding Wear (Category 1)

3.1 INTRODUCTION

The symptoms of sliding wear are often less dramatic than those of other types of wear, but its economic importance is probably greater, because it is the class of wear which generally occurs in precision machinery. It will often be the ultimate factor limiting the performance and life of such machinery. Sliding wear also occurs in a wide variety of other systems, such as wire rope over guides or pulleys, ring against ring traveller in textile machinery, high-speed sliders in teleprinters, shears and guides in metal rolling and paper-making, and so on.

Wear between sliding surfaces can involve several different mechanisms, either separately or in combination. These are adhesive wear, ploughing or cutting, and surface fatigue,* and they are all described in Chapter 1.

The effects and appearance of sliding wear can vary considerably, depending on the mechanism which is dominant and on the materials and sliding conditions. For this reason many different terms are used to describe the various forms of sliding wear, including:

Mild wear
Severe wear
Galling
Scuffing
Tearing
Wiping
Smearing
Scoring

} Adhesive (see Figs. 3.1 and 3.2)

* Strictly speaking the form of surface fatigue which is called rolling-contact fatigue can occur without any sliding, but in practical systems some sliding is always present, so that it can reasonably be included in this section.

MRW-C

(a) Steel against steel (×14)

(b) Copper against steel (×100)

FIG. 3.1 EXAMPLES OF ADHESIVE WEAR

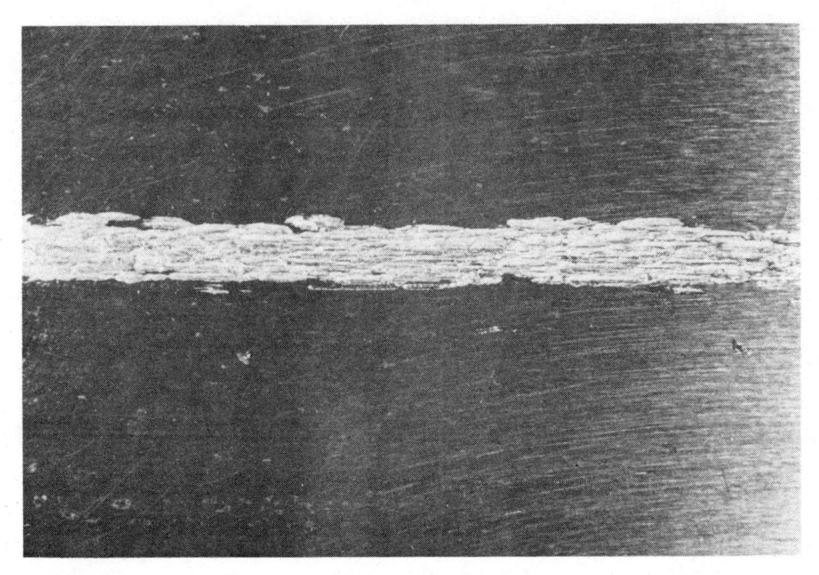

Fig. 3.2 Example of Scuffing (Severe Adhesive Wear)

Scoring* Two-body abrasion Machining wear Wire-wool failure	Ploughing or cutting
Pitting Spalling	Surface fatigue

3.2 ADHESIVE WEAR (Category 1A)

The basic mechanism of adhesive wear is described in Chapter 1 but it may be useful to summarise the various forms here.

Mild wear (Fig. 3.1(a)) takes place when protective films on the surfaces ensure that no strong adhesion takes place, so that the wear debris is fine and consists mainly of metal oxides or other reacted material rather than unreacted metal. When the protective films cease to be adequate, strong metal–metal adhesion takes place, leading to tearing, or to severe wear, characterised by coarser wear debris containing a higher proportion of unreacted metal.

* The term "scoring" is used in the United States for a severe form of the type of damage which is called "scuffing" in the United Kingdom. The same word "scoring" is often used in the United Kingdom to describe the scratching or gouging marks produced by abrasion. Because it can lead to confusion the word will not be used in this book.

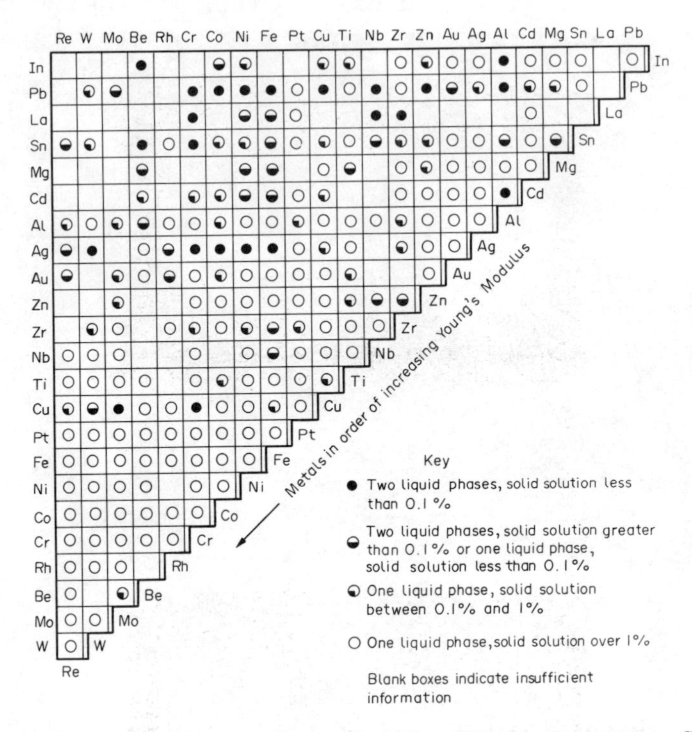

FIG. 3.3 CHOICE OF A METAL TO RESIST ADHESIVE WEAR WITH ANOTHER SPECIFIED METAL

At low speed and high load, adhesion of the asperities may take the form of galling, a type of adhesion which can be recognised by local plastic deformation or small-scale rupture of contact points.

Scuffing is a form of adhesive damage which is not clearly defined or understood (Fig. 3.2). It probably occurs with steels when the frictional heating is sufficient to permit severe adhesion and softening of one or both surfaces, but also to cause a change in alloy structure. The result is a severe visible damage mark and a change in colour of the steel, often to a silvery-white colour which may be associated with the presence of ledeburite.

One way to reduce adhesion is to use dissimilar metals, as indicated in Fig. 3.3 and Table 3.1. If one of these is significantly softer than the other, either at normal temperature or at a higher temperature caused by friction, the surface of the softer one may be wiped or smeared, as shown in Fig. 3.4. This is a process in which the surface of a softer material is carried by the counterface and transferred in the direction of sliding. It usually has a smoothed or semi-melted appearance, and it is probable that due to a combination of frictional heating and pressure there has been some melting of the surface.

Bronze Bearing ($\times\frac{1}{2}$)

FIG. 3.4 EXAMPLES OF WIPING

TABLE 3.1 COMPATIBLE PAIRS FOR SLIDING WEAR SITUATIONS

First material	Counterface materials
Steels	Copper alloys (bronze, brass), lead (white metals), tin (white metals), silver or cadmium plating, cast irons, steels of different hardness and alloy type.
Cast irons	Cast irons, copper alloys.
Copper alloys (bronze, brass)	Molybdenum coating, chromium plating, high-chrome steels, cast irons, tungsten steels.

Scuffing, wiping and smearing are probably more common in lubricated than in unlubricated systems, and represent a failure to lubricate adequately.

Some wear is probably inevitable where two surfaces slide against each other, although some scientists describe a zero-wear condition when the surfaces are fully run in. Ideally the wear should be kept in the mild wear regime. In the absence of lubricant there are several methods for reducing the risk of the more severe forms of adhesive wear, including the use of:

Harder materials.
Dissimilar materials.
Reacted surface coatings.
More readily oxidisable materials.

3.2.1 Selection of materials to resist severe adhesive wear

In order to maintain a low wear rate, the system should operate within the mild wear region. The emphasis in this section is on the sliding of steels, cast irons and copper alloys as these represent the major engineering materials. However, ceramics have been used successfully, especially at high temperatures and in cutting tool tips, but comparative test results are lacking.

3.2.1.1 STEELS

In considering the wear rate of carbon steels it is the hardness and the state of oxidation of the rubbing surfaces which are the governing factors. The bulk hardness of the material must be sufficient to support a generated oxide film and thus maintain a mild wear pattern. A bulk hardness of about 350/450 VPN is necessary with hypo-eutectoid steels, decreasing somewhat with increase in carbon content to a level of around 250 VPN for hypereutectoid steels. A maintained critical surface hardness of around 340/425 VPN generated by the rubbing itself is also effective in supporting an oxide film. With hardnesses of >700 VPN severe wear is suppressed entirely and a mild wear pattern is maintained throughout the rubbing operation, even in the absence of an oxidising medium. Therefore, to increase the wear resistance of two dry rubbing steel surfaces the components should be heat treated to obtain the necessary hardness levels, i.e. >550 VPN.

The growth of welds during sliding of carbon steels is inhibited by the inherent discontinuous structure. The hardness and composition vary throughout the surface as compared with a homogeneous austenitic stainless steel or pure iron which are more prone to large weld formation.

An increase in wear resistance of steels with ferrite–pearlite structures can be obtained by reducing the amount of free ferrite present, wear being greater for this constituent. Thus increasing the amount of carbon present, which increases the hardness as well as reducing the amount of free ferrite, is beneficial in increasing the wear resistance. This, however, does not apply to a spheroidised structure with a ferritic matrix.

The sliding wear resistance of tool steels is increased by the presence of alloy carbide, coupled with high hardness. The wear resistance of tool steels is therefore the result of chemical composition and heat treatment. The alloying elements, carbon, chromium, vanadium, niobium, tungsten and molybdenum aid wear resistance. Carbon increases hardness while the importance of the other five elements is related to their ability to form carbides. The effectiveness of Cr, W, Mo and V is in the ratio 2:5:10:40.

When considering complex alloyed materials the relationship is more complex due to other effects on the microstructure and resulting wear pattern. The main microstructural variables influencing the wear of tool steels are the presence of primary and secondary carbides, the size of the primary carbides being of particular importance—the smaller the size, the greater the wear resistance.

Heat treatment of tool steels, through the resulting variations in microstructure, can influence the wear resistance. The formation of residual austenite by the use of high hardening temperatures is undesirable and the conversion of the residual austenite to martensite does not improve wear resistance. However, on tempering the martensitic structure, precipitation of carbide takes place which significantly increases the wear resistance.

The assessment of wear resistance on the basis of hardness alone may be reasonably safe within a certain alloy system, but it cannot be applied if a radical change in the microconstituents is involved. A comparison of wear behaviour between a tool steel and a carbon-free alloy system shows that even when both have a martensitic matrix, similar hardnesses and the same amount of second phase, there is a noticeable difference in wear rate, the carbon-free alloy system being inferior to the tool steel. The difference in wear behaviour is attributed to the differences in second phase, iron carbide in tool steel and Fe_2Mo in the intermetallic alloy. Thus the hardness of a material as such is not nearly as important as some microstructural factors in determining the wear behaviour.

TABLE 3.2 STEELS FOR HIGH TEMPERATURE SLIDING WEAR

Atmosphere	Maximum continuous service temperature, °C	Steels to be used	Remarks
Oxidising	650	5–6% Cr	Difficult to weld.
	850	17% Cr (ferritic)	Difficult to weld, poor creep strength.
	900	18/8 grades (austenitic) 28% Cr (ferric)	Avoid welding, poor creep strength.
	1100	25/12, 25/20 and 35/20 (austenitic) 60% Ni alloy with 15–20% Cr	
	1150	80% Ni alloy with 15–20% Cr	
Reducing and carburising	900	25/12 and 25/20 (austenitic)	
	1000	35/20 (austenitic)	
	1100	60% Ni alloy with 15–20% Cr 80% Ni alloy with 15–20% Cr	
Reducing and sulphurous	700	18/8 (austenitic)	
	750	17% Cr (ferritic)	Avoid welding.
	900	25/12 and 25/20 (austenitic)	
	1000	28% Cr (ferritic)	Avoid welding, poor creep strength.

Particular problems arise in sliding at high temperatures, and Table 3.2 shows the temperature limits for a number of steels which can be used in sliding situations at high temperatures.

Hard wear resistant steels are often expensive to produce and also difficult to shape. The wear resistance is only required on the surface and a number of techniques have been established to provide suitable hard wear-resistant surfaces. These techniques are discussed in Chapter 10.

3.2.1.2 CAST IRONS

Grey cast irons are widely used in applications where lubrication is not feasible, because they exhibit low wear rates and can be used "sliding against themselves". During the initial sliding of graphite grey cast iron, i.e. the running-in period, wear takes place by plastic flow and fracture, with the build-up of hardened surface layers which support oxide films when formed.

Grey cast iron is section-sensitive and with variations in cooling rates, unless care is taken, the amount of free ferrite in a microstructure may vary considerably for a given composition. The presence of free ferrite markedly affects the wear pattern during running-in, the wear rate increasing with increase in free ferrite. The presence of free ferrite in the microstructure is therefore undesirable and should be controlled to less than 5%. On the completion of running-in the effect of free ferrite becomes insignificant during mild-wear sliding.

The presence of phosphorus increases the wear resistance during running-in. Optimum results are obtained with approximately 1% of phosphorus when a continuous phosphide eutectic is formed throughout the matrix, the eutectic acting as a load-bearing structure. There is, however, an additional advantage in increasing the phosphorus content still further, in that the mild/severe transition load is increased.

The resistance to wear is also influenced by the form of the graphite. A nodular form is superior to flake for irons with otherwise similar structures; the nodular form exhibits a lower running-in wear and also a lower mild wear rate for given applied conditions. Variation in the form of the flake also affects wear resistance, randomly distributed flakes (A type) producing a lower equilibrium mild wear rate than undercooled intermediate (D type), although having a greater running-in wear loss. The wear resistance of nodular iron may also be further improved by the production of a bainitic structure, by either heat treatment or casting procedure, which lowers the wear during running-in and also the equilibrium mild wear rate.

The results of a study of work-hardening grey cast irons by plastic deformation of the surface layer have indicated that the wear resistance of the irons can be considerably improved by work-hardening, although this would not normally be considered an important property of cast irons. With optimum processing conditions the surface layer can be increased in hardness by about 10%, and the surface finish is also improved. The maximum depth of hardening is about 0.5 mm and wear of the work-hardened layer is almost halved.

The wear resistance of cast iron surfaces can be improved by various surface treatments, and these are discussed in Chapter 10.

3.2.1.3 COPPER ALLOYS

Copper alloys are widely used in bearing and sliding applications. The mating material is almost always steel, which is harder than the copper alloys, and in general the design of equipment is based on wear taking place preferentially on the copper alloy. Nevertheless, it is important for the wear resistance of the copper alloy to be as high as possible.

The wear resistance of copper alloys depends upon the type of alloy, the adhesive properties of the sliding couple, structure, effects of oxide film, surface roughness and the strength and hardness of the alloy.

Dry sliding studies of copper and copper with 10% alloy additions have indicated that the friction and wear pattern is influenced by the type of alloy addition. Alloy additions of 10% of aluminium, 10% silicon, 10% tin and 10% indium were made and sliding was carried out at a speed of 300 cm/min at a load of 250 g. In all cases the alloy addition reduced both the wear rate and the friction as compared with pure copper. The reduction of wear rate was most pronounced with 10% addition of tin or silicon, and the reduction in friction with 10% aluminium or silicon. Further investigation with a ternary alloy of Cu–10% Al–5% Si gave the lowest wear rate and friction values, more effective than either single alloy addition. The low wear rates can be attributed to the stability of the formed oxides, the oxides of Al and Si having the greatest stability of the additions investigated.

(a) *Brasses*. The difference in stability of the oxide film produced during sliding by the different structures influences the wear pattern of brasses. With monophase alloys, i.e. α brass, the oxide film is extremely adherent to the metal surface and oxidative wear occurs during sliding. At low contact pressures the strong adhesion of the oxide film on the brass prevents the breaking away of large wear particles. Wear of the mating steel surface can be considerable, the wear being increased by the lapping action of iron oxide and copper oxide debris. This damage is greater if the zinc content is lower. In applications where high contact pressures prevail, the oxide film of mono-phase alloys tends to break away due to the deformation of the underlying alloy structure, and when this occurs, adhesive wear can take place. Thus, to combat this process and provide anti-wear properties under high pressure applications, the alloy requires some support to prevent deformation of the surface structure. This support is achieved by small additions of a second phase such as tin, lead and phosphorus.

The two-phase $\alpha + \beta$ brass alloys with more than 37% Zn, have a lower wear rate than the single phase alloys. Wear is of an adhesive nature giving a metallic lustre to the surface and the mating steel does not wear to any appreciable extent due to the adhesion of copper alloy particles to the steel surface.

(b) *Bronzes*. Leaded, Phosphor and Manganese bronzes generally have lower wear rates than the brasses. The improved wear rates are generally associated with the minor additions which produce a stronger supporting structure during sliding, enabling mild wear to occur by preventing break up of the surface. The addition of tin produces a hard δ phase while the addition of phosphorus produces hard particles of Cu_3P.

The addition of lead is anomalous. The decrease in wear rate is attributed to the individual lead particles found in the structure. They are able to form a surface film which at high sliding velocities provides an effective low shear strength layer.

The wear properties of copper alloys are classified in Table 3.3.

TABLE 3.3 WEAR BEHAVIOUR OF COPPER ALLOYS

Cu alloy	Wear mechanism	Wear of mating steel
60/40 type brass ($\alpha + \beta$) with some lead addition α brass plus 2% Pb	Adhesive wear	Little or no wear due to coating of copper alloy particles on the steel surface.
Mn bronze, Al bronze. Tin bronze with high Pb	Mild wear	Mating steel scarcely wears.
70/30 brass	Mild wear	Tendency to seizure: restricted mainly to lubricated systems.

3.2.1.4 OTHER METALS

The majority of sliding surfaces in engineering will be of steel, cast iron, brass or bronze. White metals are used almost entirely in lubricated systems, where their softness enables them to take up an appropriate shape during running in to provide good conformity to a counterface, and where to some extent they act as sacrificial surfaces to avoid damage to a more expensive counterface.

A few other metals, such as chromium, molybdenum and nickel, are sometimes used as sliding surfaces to provide special behaviour against difficult counterface materials or in corrosive conditions, but they are normally employed as a surfacing material on a steel substrate (see Chapter 10).

3.3 PLOUGHING OR CUTTING

Basically, ploughing or cutting wear occurs when a hard asperity ploughs or cuts through a softer material. An extreme example is shown in Fig. 3.5. The techniques for avoiding it are therefore to minimise the presence of asperities on the harder material and to increase the hardness of the softer material.

There are two general situations in which a hard surface is designed to slide against a softer surface. One is in lubricated systems, where the less critical component is often made of a soft metal or non-metal in order to improve running-in, to embed any deleterious particles which may be present in the lubricant, and to minimise wear of the more critical component. The other situation is in the increasing use of a soft polymer sliding against a hard metal.

There is some evidence that the wear rate of the softer material is inversely proportional to its hardness, although it is doubtful if this simple relationship always applies. However, the wear rate of the softer material generally decreases as its hardness increases towards that of the harder surface. The selection of the hardness of the softer material will therefore be a compromise. Softness is desirable for the purposes listed in the last paragraph, but the hardness should not be reduced further than is needed to achieve those purposes.

Within normal engineering limits the harder surface should be as smooth as possible in order to minimise wear of the softer surface. This applies down to finishes of the order of 0.1 μm (4 μin) C.L.A. Where a polymer is being used against a harder metal, it has even been found beneficial to incorporate a mild abrasive filler in the polymer to produce the optimum finish on the metal surface.

3.4 SURFACE FATIGUE

The mechanism of surface fatigue is described in Chapter 1. Because it is initiated by subsurface micro-cracks, it follows that it will be encouraged by any factors which encourage cracking. These include:

(i) Hard surface coatings, especially if there is a marked (i.e. non-diffusive) interface between coating and substrate.
(ii) Case-hardening.
(iii) Inclusions.

Alloy Steel Shaft (×1)

FIG. 3.5 EXAMPLE OF CUTTING OR MACHINING WEAR

(iv) A coarse grain structure.
(v) Brittleness generally, as in ceramics or cast irons.

Surface fatigue tends to arise where there are high local stresses, especially in non-conformal contacts such as those in gears and rolling bearings. High strength is therefore necessary, but this must be combined with toughness. For rolling bearings a 1% chrome steel of the EN31 or AISI 52100 type, hardened to about 800 VPN, is almost universally used.

3.5 GUIDELINES TO MATERIAL SELECTION

Because adhesive wear is complex, a key to the selection of materials to reduce adhesive wear is provided in Fig. 3.6. Table 3.4 gives general guidelines to selection, while Tables 3.5 to 3.7 give guidelines to compositional, structural and hardness factors and the use of surface treatments or coatings.

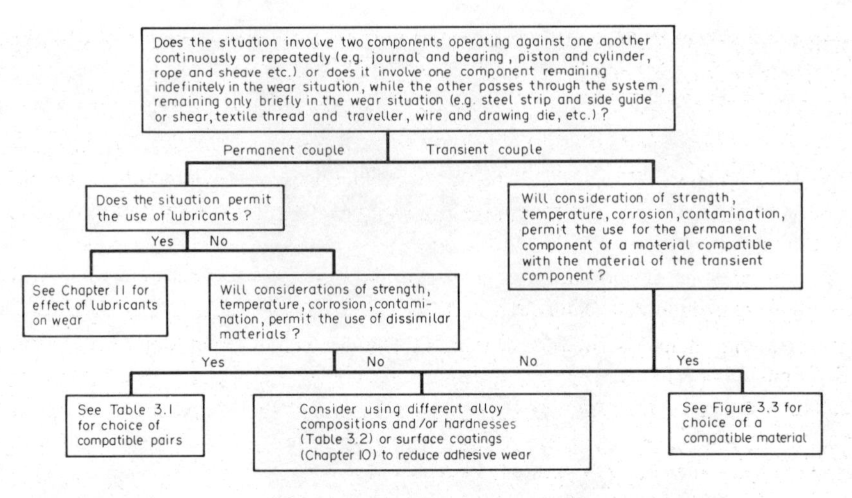

FIG. 3.6 KEY TO THE SELECTION OF MATERIALS TO REDUCE ADHESIVE WEAR

TABLE 3.4 GENERAL GUIDELINES FOR THE SELECTION OF MATERIALS TO RESIST SLIDING WEAR

General
1. The worst type of sliding wear is severe adhesive wear, and materials should be selected to avoid it.
2. As far as possible use different materials in contact. See Table 3.1 for compatible materials.
3. Consider using a softer material for the component which is most easily replaced, so that wear occurs preferentially on that component.
4. Where a hard material runs against a soft one, the hard surface should have a smooth finish (e.g. better than 0.2 μm).
5. If circumstances dictate the use of similar metals in sliding contact, use compositional and structural factors to reduce wear (see Tables 3.5 and 3.6).
6. There is a transition from mild wear to severe adhesive wear as temperature increases. The combination of contact load and speed must be kept below those which cause severe wear. (At even higher temperatures a system will revert to mild wear, but it is not usually possible in practice to make use of this second transition.)
7. A good surface oxide film will inhibit severe wear as long as the substrate hardness is high enough to prevent break-up of the film under load.

TABLE 3.5 GUIDELINES TO COMPOSITIONAL FACTORS IN THE SELECTION OF MATERIALS TO RESIST SLIDING WEAR

Composition factors
8. The tendency for severe adhesion is reduced by using mutually insoluble materials (Fig. 3.3) With steels, metals of the B sub-group of the Periodic Table are beneficial.
9. Metals which resist oxidation are more likely to experience severe wear with tearing and metal transfer.
10. Alloying elements such as carbon, chromium, vanadium, niobium, tungsten and molybdenum assist wear-resistance of steels.
11. If a steel has a high chromium content in relation to its carbon content, it will be prone to adhesive wear. Wear-resistance will be improved if the chromium is in the form of primary or secondary carbides.
12. The wear-resistance of an alloy steel is developed in the order vanadium > molybdenum > tungsten > chromium, and this is related to their ability to form carbides.
13. The adhesive wear-resistance of steel improves as the free ferrite content decreases. Free ferrite should be below 5%.
14. Presence of up to 3% phosphorus improves wear-resistance of grey cast irons.

TABLE 3.6 GUIDELINES TO STRUCTURAL FACTORS IN THE SELECTION OF MATERIALS TO RESIST SLIDING WEAR

Microstructural factors
15. The smaller the grain size the better is the wear-resistance.
16. A discontinuous structure helps to prevent severe welding between steels (e.g. carbon steels are better than homogeneous austenitic stainless steels or pure irons).
17. The type of crystal structure influences sliding wear. A hexagonal crystal structure maintained over the full operating range is beneficial.

TABLE 3.7 GUIDELINES TO HARDNESS AND SURFACE TREATMENTS IN THE SELECTION OF MATERIALS TO RESIST SLIDING WEAR

Hardness

18. For steel a hardness greater than 700 VPN suppresses severe wear entirely.
19. To improve adhesive wear-resistance of two rubbing steel surfaces they should preferably be between 550 and 750 VPN.
20. The critical hardness level for steels decreases with increase in carbon content. For hypoeutectoid steels (<0.85% C) the hardness should be 400 VPN or more. For hypereutectoid steels (>0.85% C) the hardness should be greater than 250 VPN.

Surface treatments or coatings

21. Where structural requirements prevent the use of suitable metals in sliding contact, the use of surface treatments or coatings may be beneficial (see Chapter 10).
22. A suitable material may be applied as a coating or facing on a less suitable material.
23. A surface treatment which increases surface hardness may be beneficial (e.g. carburising or nitriding) but the depth of the treatment must be related to the maximum acceptable wear depth.
24. Surface treatments which produce chemical changes in the surface (e.g. phosphating, sulphidising) may be beneficial, especially in running in new components.

Chapter 4

Fretting (Category 2)

4.1 INTRODUCTION

Fretting is a form of surface damage which occurs when surfaces which are loaded against each other undergo small-amplitude oscillatory slip. It is also described by a number of other names, including fretting corrosion, fretting oxidation, or false brinelling. The damage takes the form of surface pitting accompanied by loose debris, and can lead to fatigue cracking at stresses below the normal minimum.

Fretting is a common problem in all sorts of engineering environments, and has at times resulted in very expensive damage to equipment. There are two general situations in which it can arise. The first is in systems where the two components are designed to move relative to one another, such as in bearings, gears, valves, and especially splines, but where under certain service conditions the relative movement is of very small amplitude and reciprocating. Under these circumstances fretting damage may be superimposed on some other form of wear which occurs during gross relative movement, and the fretting debris can cause seizure of the mechanism. This can be especially dangerous if fretting causes seizure of a safety valve or other safety device. An example is shown in Fig. 4.1(b)

The second fretting situation occurs with components which are not designed to move relative to each other, such as in bolted or rivetted joints, clamps, press fits or even interference fits, where a superimposed stress or vibration is sufficient to produce small-amplitude relative movement. Fretting can be difficult to detect in such situations, but can cause loss of stiffness in the assembly or difficulties in disassembly. An example is shown in Fig. 4.1(a).

4.2 MECHANISM OF FRETTING

There is no general agreement about the mechanism of fretting. The most widely accepted theory is that adhesive asperity contacts form between the two surfaces in contact under load. When relative

MRW-D

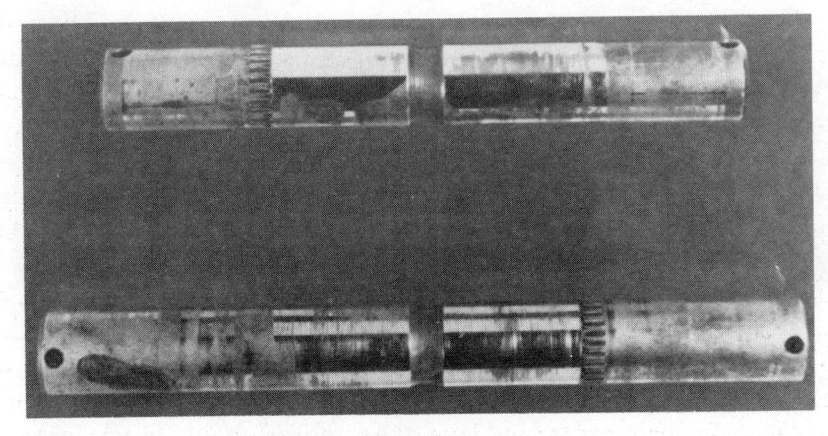

(a) Hardened Steel Pin (×1)

(b) Alloy Steel Inner Race (×$\frac{1}{4}$)

FIG. 4.1 EXAMPLES OF FRETTING

movement takes place between the surfaces, the junctions fracture, and ultimately loose particles are formed by the breaking away of an asperity from a surface. The loose particle is fully oxidised either before or after detachment from the surface. Because the relative motion is a low-amplitude oscillation, loose particles are trapped between the two surfaces and cause increased wear of the surfaces by abrasion.

There are several objections to this proposed mechanism. The first is that on the basis of contact load and total sliding distance the wear rates are of the same order as those found in mild sliding wear, so that there is no apparent need to postulate any special accelerated wear mechanism. The second is that normally metallic oxides are not abrasive to their own base metals. A third objection is that even at the higher amplitudes there is little or no sign of any directionality in the surface damage which would support the theory of abrasion.

An alternative possibility is that fretting is quite simply a form of surface corrosion fatigue, in which oscillating strains in the presence of a corrodent (oxygen) lead to surface fatigue cracks which propagate to the stage where a particle is loosened from the surface. This mechanism accounts for the lack of any directionality in the surface damage and for the fact that the surfaces of the pits are oxidised or corroded even when freshly formed.

The presence of a corrosive substance such as oxygen is necessary for fretting to occur, but the corrodent is not necessarily oxygen.

4.3 NATURE OF THE DAMAGE

Fretting is characterised by pitting of the surfaces and the formation of loose debris. The pits vary in size from 0.1 mm (0.004 in) to 2.5 mm (0.1 in) across, and may overlap to give continuous areas of damage. The surfaces of the pits are always fully oxidised or otherwise corroded, and are rough on a microscopic scale. This enables them to be distinguished from cavitation erosion (rough but unoxidised) or electrical pitting (smooth and unoxidised) but confusion is possible in the early stages with certain types of corrosion.

The debris is fine and almost fully oxidised or corroded (i.e. with little or no unreacted metal present). With ferrous metals it consists largely of ferric oxide, Fe_2O_3, and has a characteristic "cocoa" appearance while with aluminium alloys it tends to be black. It may be present in copious quantities, and under some circumstances the first indication of fretting may be the appearance of the fine debris exuding from the contact zone.

There are basically four hazards arising from fretting. Where it arises in bearings or gears it will cause rough running, and the resulting dynamic stresses can cause further damage. In a bolted, rivetted or tapered joint it can lead to looseness in the joint, and similarly in a spline it can cause backlash in the drive.

When fretting takes place in a press fit, such as between the outer surface of a rolling bearing and the inner surface of the housing, it can cause loss of fit, and gross movement. Alternatively the generation of a rough fretted surface may cause the bearing to seize in the housing, so that when the time comes for a change, the bearing cannot be withdrawn from the housing. A similar situation can arise with a shrunk fit, but usually the higher contact pressure in a shrunk fit will prevent fretting. This type of seizure can also arise in safety valves and safety switches which are normally static, and the result can be a failure of the safety device to operate when required.

A complete structural failure of equipment can arise when fretting initiates fatigue failures, and this can arise at stresses below the normal critical fatigue stress. This provides some support for the theory that fretting is itself a stress corrosion fatigue process. Splines sometimes fail by this mechanism, when fretting causes backlash which leads to dynamic loading and the spline teeth fail by fatigue at relatively low stresses.

4.4 CONTROL OF FRETTING

4.4.1 Changing the amplitude of the oscillation

In a given system fretting will usually only take place within a relatively narrow range of oscillation amplitudes, and it can be reduced or eliminated by changing the amplitude.

Where the system involves a contact which is designed to move, such as a bearing or gear, it may be difficult to reduce the amplitude, and the best solution may be to increase it.

In a gear coupling, for example, fretting can sometimes be eliminated by increasing the degree of misalignment. In a bearing it may be eliminated by superimposing a steady rotation. In a servo-valve or similar control mechanism it may be improved by deliberately designing the system to "hunt" over a greater amplitude. An exception is a safety valve or switch for which it may be better to try to reduce oscillation by increasing contact pressure.

The most general technique for reducing oscillation is to increase contact pressure. This can be done by increasing the contact load (e.g. in a bolted joint) or by reducing the contact area (e.g. in a bearing housing), by milling recesses in the contact surface. It should be borne in mind, however, that in some systems it is the frictional force which limits the amplitude, and this is determined by the contact load and not the contact pressure. It seems possible that under certain circumstances improved lubrication could give an increased amplitude of oscillation, but there is no real evidence that this is effective.

4.4.2 Reducing the availability of oxygen/corrodent

Since fretting depends on oxidation or corrosion, it follows that eliminating oxidation or corrosion will eliminate fretting. In practice it is usually not possible to eliminate the corrodent, but by reducing its access to the surface, fretting can be reduced. This is done by using paints or rubber seals, or by means of viscous oils, especially if they contain anti-oxidants.

Bonded molybdenum disulphide coatings have been shown to delay the occurrence of fretting, and molybdenum disulphide has been incorporated in paints and jointing compounds to prevent fretting of bolted joints (see Table 4.1).

TABLE 4.1 RECOMMENDED COATINGS TO REDUCE FRETTING

Substrates	Coating	Typical application
Steels	Paints	Inspection covers.
	Bonded MoS_2 films	Rivetted or clamped joints.
	Copper plating	Splines, electrical contacts.
	Cadmium plating	Water pump supports.
	Silver plating	Electrical contacts.
	Phosphate Sulphide	Fuel or oil pump components, clamped or bolted components in oil or grease.
Copper alloys	Paints Cadmium plating	Mountings for guides or large electrical contacts.
Aluminium alloys	Paints Anodise	Airframe structural joints.
Titanium alloys	Heavy anodise Molybdenum	Airframe joints.

Note: Where similar substrate metals are used for both components, only one should be coated, or dissimilar coatings should be used

4.4.3 Metal selection

The use of soft metal coatings such as copper, aluminium or cadmium can prevent fretting, either by absorbing the relative movement without fatigue or by changing the oxidation behaviour. The use of unlike metals in contact is recommended to reduce the initial tendency to adhesive wear (see Table 4.2). Complete separation of the surfaces by non-metallic materials will also often eliminate the problem (see Tables 4.1 and 4.3).

TABLE 4.2 RECOMMENDED COMBINATIONS OF MATERIALS TO REDUCE FRETTING

Primary material	Secondary material	Application
Mild steel	Soft coatings	Bolted joints.
Alloy steels	Silver plating	Bolted jointed, safety switches.
Carburised steel	Copper alloys	Shim in bearing housings.
Copper alloys	Steels	Bearing housings.
Aluminium alloys	Steels, molybdenum	Bolted or rivetted joints.

Note: These combinations will reduce the incidence of fretting but are not likely to eliminate it.

TABLE 4.3 RECOMMENDED INSERTS OR INTERLAYERS TO REDUCE FRETTING

Material of insert	Typical application
1. Copper shim between steels	Clamped joints.
2. Rubber sheet, especially nitrile or PVC	Vehicle door frames, inspection covers.
3. PTFE sheet	Pressure covers, hydraulic joints.
4. Nylon	Pipework, structural joints.

Note: By combining damping with separation, inserts may completely eliminate fretting.

4.4.4. Design to reduce fatigue fractures

Conventional design features to eliminate stress concentrations, such as radiusing of edges, will reduce the tendency for fatigue fractures to take place if fretting occurs.

General guidlines for fretting are given in Table 4.4

TABLE 4.4 GENERAL GUIDELINES FOR THE SELECTION OF MATERIALS AND CONDITIONS TO REDUCE FRETTING

1. Reduce or eliminate relative oscillation if possible.
2. Try increasing amplitude of relative oscillation.
3. Use compatible dissimilar metals (see Table 3.1).
4. Coat one surface with a soft metal or non-metal (see Table 4.1).
5. Coat both surfaces with dissimilar coatings (see Table 4.1).
6. Interpose soft metal or non-metallic insert between surfaces (see Table 4.3).
7. If lubrication is desired, use bonded molybdenum disulphide film or viscous oil with anti-oxidants.
8. Eliminate points of high stress concentration such as sharp steps or edges by radiusing, etc., to reduce risk of fatigue failure.

Chapter 5

Three-body Abrasion (Category 3)

5.1 INTRODUCTION

"Three-body abrasion", sometimes known as "grinding wear", manifests itself in the comminuting or grinding industries where wear of ball mill liners and grinding balls can be a costly overhead. In general it occurs under conditions of low load and high stress, and the wear situation is governed by two principal factors, the properties of the abrasive particles and those of the materials against which the abrasive particles are rubbing.

In most situations the particle properties are fixed with respect to hardness and toughness, and can therefore be regarded as a constant of the system. Thus it is the properties of the wearing metallic surfaces which must be considered in order to minimise or control the degree of wear occurring.

An example of three-body abrasion is shown in Fig. 5.1.

5.2 NATURE OF THE WEAR SITUATION

The physical situation consists of two surfaces in relative motion with abrasive particles between them transmitting load from one surface to the other. The local stresses at the points of contact are likely to be very high due to the actual load supporting area, the particles, being considerably less than the area of load application. The total real area of contact in the system will be very much less than the total area of the two surfaces, so that the mean pressure over the whole surface will be relatively low. Limiting factors of the system are therefore the crushing strength of the particles and the yield strength of the surfaces. The compressive crushing strength of most minerals is relatively low, e.g. for quartz ~30 000 psi and therefore the particles are broken down fairly easily. The broken particles are, however, sharp and can cause deterioration of the metallic surfaces

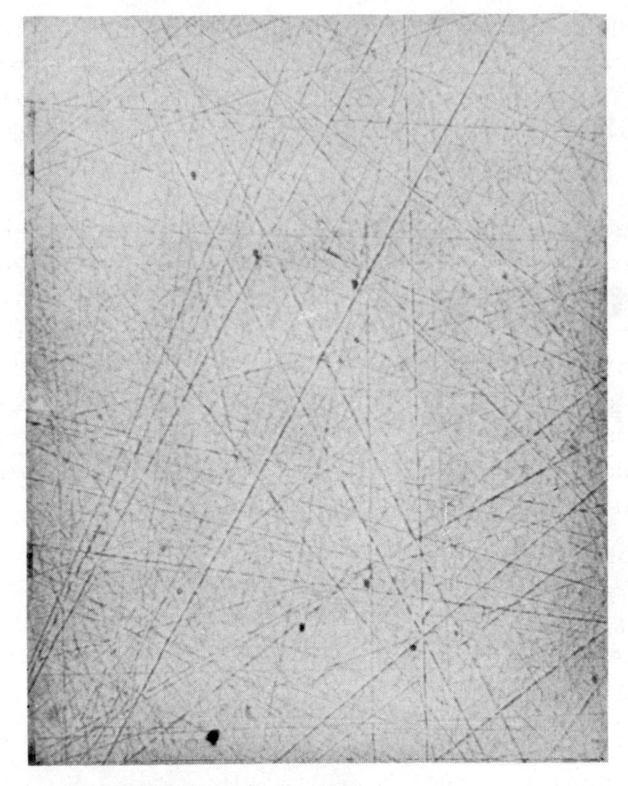

FIG. 5.1 EXAMPLE OF THREE-BODY ABRASION

by scratching, plastic flow and microcracking. The surfaces must therefore be hard enough to combat the deterioration.

However, the surfaces are also in practice the surfaces of engineering components subjected to structural, impact or bending loads, and therefore the structural integrity of the components will also be of importance. It follows that although hardness is the dominant requirement for the surfaces, there will also be a need for toughness and structural strength.

5.3 REQUIREMENTS FOR WEAR RESISTANCE

The hardness ratio of the abrading particles and the surfaces or the hardness of the individual structure components of the surface material, is of critical importance. Generally for a material to be appreciably worn, the abrasive must be harder than it, and a marked drop in the wear rate occurs when the hardness of the abrasive mineral is less than that of the material which it is abrading. The hardness of the surface is also important in its capability of blunting the sharpness of the cutting points on an abrasive particle. Harder alloys tend to increase fracture of the cutting points and reduce the wear rate. The intrinsic toughness of the abrading particles is important in grinding operations due to the ability of tough particles to withstand fracture. Since wear rate decreases with decrease in particle size it is important to overcome particle toughness by using hard alloys which can resist and fracture the particles.

The structure of the alloy in addition to the hardness is important when assessing wear-resistance. Materials which contain hard phases, e.g. hypereutectoid steels and many cast alloys and irons are very sensitive to the scale of shattering at the abrasive surface. The contribution of the hard constituents to wear-resistance depends upon their size relative to that of a chip cut by the abrasive. Small particles are dug out while large ones have to be worn and therefore contribute enormously to the wear-resistance. Steel matrices in the order of resistance to three-body abrasion are:

Increasing wear-resistance ↑

High carbon martensitic matrix.
High carbon pearlitic matrix.
Bainite.
Soft pearlite.
Pearlite and ferrite.
Low carbon ferrite.

The relative resistance of some classes of steel and cast iron is shown in Table 5.1, and it can be seen that the wear-resistance is not

TABLE 5.1 RELATIVE RESISTANCE OF SOME GROUPS OF METALS TO THREE-BODY ABRASION

Group	Material	Hardness VPN 30 kg	Relative wear rate
1	High chromium white cast irons	865	89
2	Martensitic 1% carbon Cr–Mo steel	595	100 (std)
3	"Martensitic" Ni–Cr–Mo white iron, chill cast	674	107
4	Martensitic 0.7% carbon Cr–Mo steel	653	111
5	Nickel–chromium alloyed martensitic white cast irons	595	116
6	Martensitic 0.4% carbon Cr–Mo steel	595	120
7	Pearlitic 0.8% carbon Cr–Mo steel	382	127
8	Austenitic manganese steel	484	195
9	Pearlitic low Cr white iron	484	195
10	Unalloyed steels	120	225

(Increasing wear-resistance ↑ from group 10 to group 1)

Note: Hardness values determined after service in a ball mill.

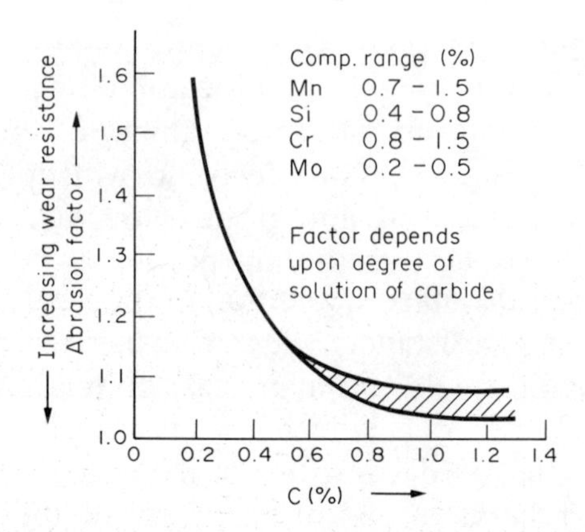

FIG. 5.2 EFFECT OF CARBON CONTENT ON WEAR-RESISTANCE OF MARTENSITIC CAST STEELS (TESTED IN GRINDING A SILICEOUS ORE)

related directly to hardness. This is shown by the three classes of martensitic chromium–molybdenum steels, for which a closer relationship occurs with carbon content than with hardness. A similar effect is seen in Fig. 5.2, which shows the variation of abrasion resistance with carbon content for a series of martensitic cast steels.

Table 5.2 lists some martensitic chromium–molybdenum steels, and Tables 5.3–5.8 some irons and steels in the other groups in Table 5.1.

The effect of microstructure on three-body abrasion resistance is complex, and some examples are shown in Table 5.9.

In some situations ceramics may be usable to resist three-body abrasion. Their advantages and disadvantages are listed in Table 5.10, and some which have been used successfully are shown in Table 5.11.

General guidelines to the selection of materials to resist three-body abrasion are presented in Table 5.12, and guidelines to compositional and microstructural factors in Table 5.13.

Finally, the effect of abrasive particle properties on selection is shown in Table 5.14.

TABLE 5.2 MARTENSITIC CHROMIUM–MOLYBDENUM STEELS

Type	% Composition					Heat treatment	Applications
	C	Mn	Si	Cr	Mo		
1	0.7	0.9	0.7	1.3	0.4	Hardened in hot bath	Liner plates, other wear-resistant coatings.
2	1.0	0.7	0.6	6.0	1.0	Air-hardened	Large grinding balls, e.g. 125 mm diameter.
3	0.8	0.6	0.4	0.4	—	Water-hardened	Smaller grinding balls, 50–90 mm diameter.
4	1.2	0.7	0.6	1.0	0.3	Oil-hardened	
5	0.5	0.7	0.6	0.9	0.5	Oil-hardened	Mill liners, gyratory crushers.

Note: Type 5, oil quenched and tempered, hardness exceeds 500 Hv.

TABLE 5.3 HIGH CHROMIUM WHITE CAST IRONS

Type	Typical composition % C	Mn	Si	Cr	Mo	Ni	Heat treatment	Applications
15–3	2.4–3.6	0.7	0.6	15	3	—	Air-hardened	High wear situations: impact mills: ore grinding.
15–2–1	3.4	0.7	0.6	15	2	1.0	Air-hardened	More suitable than 15–3 for heavy sections.
20–2	3.0	0.7	0.6	20	1.5	0.5	Air-hardened	
12Cr–Mo	3.2	0.7	0.6	12	0.5	—	Oil-hardened	Thin sections, e.g. small grinding balls.
25–32Cr	2.3–3.0	1.0	1.0	27	0.5	—	Air-hardened	Sand & ballast pumps in processing of ceramics.
8Cr–6Ni	3.3	0.6	1.5	8	—	6.0	Air-hardened	Heavy sections for medium wear-resistance.
10–15Cr	1.2–2.5	1.5	0.6	12	—	—	Air-hardened	

Notes:
1. *Characteristic structure*—primary and secondary carbides, generally of the M_7C_3 type, surrounded by an austenitic or martensitic matrix. Mo added to suppress pearlite formation.
2. *Heat treatment*—favourable response. Suitable annealing gives improved machinability and also martensitic hardening can be obtained by quenching in oil, air or salt bath.
3. 25–32Cr—Good combination of wear-resistance, ductility, lower cost.
4. 8Cr–6Ni—Ni added to suppress pearlite formation in heavy sections.
5. 10–15Cr—Economic but less wear-resistant.

TABLE 5.4 NICKEL–CHROMIUM ALLOYED "MARTENSITIC" WHITE CAST IRONS (NI–HARD)

Type	Composition %							Hardness VPN	Izod Impact (ft lbs)	Applications
	C	Si	Mn	S	P	Ni	Ci			
Type 1 Standard	3.0/3.6	0.4/0.7	0.4/0.7	0.15 max.	0.40 max.	3.5/4.75	1.4/2.5	585 min.	20–30	Pump casings and impellers for abrasive slurries: mill liners: grinding balls: drop balls.
Type 2 Hi-strength	2.9 max.	0.4/0.7	0.4/0.7	0.15 max.	0.40 max.	3.5/4.75	1.4/2.5	560 min.		
Type 3 Low carbon	1.3	0.6	0.5	0.1 max.	0.1 max.	4.5	1.6	400–450		
Type 4 Eutectic	2.6/3.2	1.7/2.0	0.4/0.6	0.1 max.	0.06 max.	5.0/5.5	7.5/9.0	640 min.	40–50	
Added molybdenum (1.8–4.0%)								690–800	40–50	

TABLE 5.5 PEARLITIC 0.8% CARBON CR–MO STEEL

Type	Composition %						Normal casting treatment	Hardness attainable VPN	Izod impact (ft lbs)	Applications
	C	Mn	Si	Cr	Mo	Ni				
1	0.9	0.7	0.5	2.2	0.4	—	Air-hardened	300–450	120–90	Low impact stressing liners, sections of 50–150 mm.
2	0.8	0.8	0.5	1.7	0.3	0.8	Air-hardened	300–450	120–90	
3	1.1	0.7	0.6	1.0	0.2	—	Air-hardened from casting heat	300–450	120–90	High stress grinding balls 50–125 mm diameter.
4	0.8	0.7	0.6	1.0	0.2	—	Air-hardened from casting heat	300–450	120–90	

Notes:
1. These steels can be obtained in as-cast, air-hardened or chill-cast condition.
2. When quenched and tempered, martensitic structures are produced, giving excellent hardness and toughness, comparable with some in Table 5.4.

TABLE 5.6 AUSTENITIC MANGANESE STEELS

Material	% Element C	Mn	Si	Cr	Mo	Ni	Strength psi × 10^3	BHN VPN 30 kg	Heat treatment	Applications
A	1.05–1.35	11.0 min	0.3–1.0	—	—	—				
B-1	0.9–1.05	11.5–14.0	0.3–1.0	—	—	—				
B-2	1.05–1.2	11.5–14.0	0.3–1.0	—	—	—	45–55	190–210	Water quenched from 1050°C	Liners in gyratory and conical breakers and crushers. B grades used for large section castings, or where subject to thermal embrittlement.
B-3	1.12–1.28	11.5–14.0	0.3–1.0	—	—	—				
B-4	1.2–1.35	11.5–14.0	0.3–1.0	—	—	—				
C	1.1–1.3	12.0–14.0	0.3–1.0	1.5–2.5	—	—	50–60	200–335	Water quenched from 1100°C	Digger teeth: shell & bowl liners.
D	1.05–1.15	12.0–14.5	0.5–0.1	—	—	3.5–5.0	40–55	160–190	Water quenched from 1050°C	Welding.
E-1	1.0–1.2	13.0–14.0	0.4–0.6	—	0.9–1.1	—	52–62	180–210	Water quenched from 1050°C	Large crusher linings.
E-2	1.0–1.2	13.0–14.0	0.4–0.6	—	1.8–2.2	—	60–70	180–210	12 hrs 600°C Water quenched from 1000°C	Longer life, heavy duty crushers.

Notes:
1. These steels have the property of work-hardening under repeated heavy impact to as much as 400/500 VPN, giving good resistance to grinding abrasion.
2. With C material care must be taken in controlling % C and heat treatment to obtain proper microstructure and physical properties, to avoid embrittlement.
3. In D material the nickel reduces tendency to embrittlement if overheated in welding.
4. E-2 material is costly, and application so far is limited.
5. Copper may be added to give improved strength, toughness and work-hardening.
6. Titanium sometimes added as a grain refiner.

TABLE 5.7 PEARLITIC LOW CHROMIUM WHITE IRONS

Composition %							
C	Si	Mn	S	P	Cr	VPN	Applications
3.3/3.6	0.4/1.0	0.5/0.7	0.15	0.30	—	420	Grinding mill liners
3.3/3.6	0.4/1.0	0.5/0.7	0.15	0.30	1.0/2.0	480	(small diameter mills).

Notes:
1. Produced in the chill cast condition.
2. Hardness increases with % C.
3. Addition of Cr controls chill depth and ensures graphite-free structure.
4. Fair wear-resistance under low impact conditions.
5. Economic to produce.

TABLE 5.8 UNALLOYED STEELS—TYPICAL COMPOSITIONS AND PROPERTIES

Type	Chemical composition % C	Si	Mn	S	P	Condition	Mechanical properties UTS T/″	Izod ft/lbs min	VPN
"40" carbon (En8 series)	0.35–0.45	0.05–0.35	0.60–1.00	0.05 max.	0.05 max.	Normalised hardened and tempered.	35 40	10 25	150/210 180/250
"50" carbon (En43 series)	0.45–0.55	0.05–0.35	0.70–1.00	0.06 max.	0.06 max.	Normalised hardened and tempered.	40 45–50	15 25	180/250 200/280
"55" carbon (En9 series)	0.50–0.60	0.05–0.35	0.50–0.80	0.06 max.	0.06 max.	Normalised hardened and tempered.	45 50–55		200/270 230/320
"60" carbon (En43 series)	0.60–0.65	0.05–0.35	0.40–0.60	0.06 max.	0.06 max.	Produced to specified composition limits only.	—	—	—
Higher carbon (En42 series)	0.75–0.90	0.35 max.	0.60–0.90	0.05 max.	0.05 max.		—	—	—

Notes:

1. Whole series is heat treatable to produce range of hardness and toughness, more expensive.
   (a) 0.3–0.6% C, varying quenching rates followed by tempering produces hardness range 220–450 VPN. Applicable to crankshafts and couplings.
   (b) 0.6–0.9% C, varying quenching rates followed by tempering produces hardness range up to 600 VPN. Applicable to agricultural tools.
2. Application of surface hardening treatments, e.g. carburising flame or induction hardening, gives surfaces of moderate wear-resistance at lower cost than in Note 1.

TABLE 5.9 EFFECT OF MICROSTRUCTURE OF THREE-BODY ABRASION RESISTANCE

Metals	VPN 30 kg	Influence of microstructure
SG Iron	270	Free graphite has deleterious effect.
SAE 1020 steel	137	Wear-resistance dependent upon pearlite content in untreated state.
Pearlite white cast iron	425	Hard, brittle carbides in soft matrix, broken up by grinding action.
Austenitic manganese steel	212	Austenitic grains, mechanical properties impaired if some cementite precipitation occurs at grain boundaries.
Chrome steel C.R.O.	283	Increasing abrasion resistance with increase in pearlite with % C.
Low alloy martensitic steel	510	Martensitic structures give better abrasion resistance.
Low C chrome moly steel	358	Abrasion resistance results from production of fully martensitic structure.
Eutectic Ni-hard	640	Normally contain some retained austenite in matrix, thus slightly inferior to high-chromium martensitic white irons.
Regular Ni-hard	570	
High Cr Martensitic white iron	740	Very hard carbides in predominantly martensitic matrix beneficial.

Increasing wear resistance ↓

Note: The very hard but expensive high chromium white cast iron may be more economical in the long term because it has a much longer life than the softer alloys.

TABLE 5.10 SOME ADVANTAGES AND DISADVANTAGES OF CERAMICS FOR RESISTANCE TO THREE-BODY ABRASION

Advantages	Disadvantages
1. Highest available resistance to certain types of wear.	1. Relatively poor resistance to mechanical or thermal shock.
2. Resistant to chemical attack, and therefore to corrosive wear.	2. Size of components limited.
3. Avoid contamination by metallic wear debris in ore grinding.	3. Difficult to fabricate to tight tolerances.
4. Best for grinding rather than crushing.	4. High cost.

TABLE 5.11 CERAMICS FOR RESISTANCE TO THREE-BODY ABRASION

Material	Strength		Maximum component weight kg	Maximum component length, m	Applications
	Rupture modulus	Compressive strength			
Dense alumina	56 000 psi	280×10^3 psi	15	0.3	Small hydrocyclones: ball mills: scrapers.
Self-bonded silicon carbide	22 800 psi	150×10^3 psi	25	1.0	Cyclone separators: scrapers.
Fused cast zirconia–alumina	1500 psi	35×10^3 psi	150	1.0	Ball mills: surface paving.
Silicon nitride-bonded silicon carbide	7000 psi	20×10^3 psi	70	1.0	Hydrocyclones: slurry pumps.
Dense porcelain	15 000 psi	100×10^3 psi	80	1.5	Slurry pumps: ball mills.
Fused cast basalt	3000 psi	70×10^3 psi	15	0.25	Sinter decks.

↑ Increasing abrasion resistance (from Fused cast basalt up to Dense alumina)

Note: Other ceramics such as titanium nitride and sialons have been successfully used, but no useful comparative wear test results have been published.

TABLE 5.12 GENERAL GUIDELINES TO THE SELECTION OF MATERIALS TO RESIST THREE-BODY ABRASION

1. The dominant requirement for resistance to three-body abrasion is hardness.
2. For technically pure metals wear-resistance is proportional to hardness in the fully annealed condition.
3. For most practical metals the wear-resistance is controlled by work-hardening and does not vary greatly with initial hardness.
4. Hardness is only effective in increasing abrasion resistance if the value is greater than about half that of the abrasive.
5. Little or no further abrasion resistance is obtained by an increase in hardness beyond 1.3 times the hardness of the abrasive.
6. Surface treatments or coatings can be used to provide improved abrasion resistance, but the depth must be related to the permissible wear depth (see Chapter 10.

TABLE 5.13 GUIDELINES TO COMPOSITIONAL AND MICROSTRUCTURAL FACTORS IN THE SELECTION OF MATERIALS TO RESIST THREE-BODY ABRASION

1. The best structure for carbon steels appears to be lower bainite.
2. The abrasion resistance of pearlitic steels generally increases with the addition of carbide-forming elements.
3. For alloy steels the form of the carbide is the most important factor. Removal of primary carbides and precipitation of fine uniform secondary carbide in the matrix is beneficial.
4. For alloy steels the best abrasive wear-resistance is obtained with uniformly distributed fine carbide in a martensitic matrix with some residual austenite.
5. Vanadium and niobium carbides give the best resistance to abrasive penetration of alloy steel surfaces.
6. Martempering gives high wear-resistance but leaves residual micro-stresses.

TABLE 5.14 GUIDELINES TO THE EFFECT OF ABRASIVE PARTICLE PROPERTIES ON THE SELECTION OF MATERIALS TO RESIST THREE-BODY ABRASION

1. If the abrasive particles are fine the presence of coarse hard particles in an alloy can give considerable improvement in abrasion resistance.
2. If the abrasive is coarse a more homogeneous alloy structure is desirable.
3. If the abrasive is fine a brittle metal or ceramic can be used, but with coarse abrasives surface cracking can lead to increased wear.

Chapter 6

Gouging Wear (Category 4)

6.1 INTRODUCTION

When abrasive lumps or particles rub against a surface with sufficient force to gouge out material, this is known as gouging wear or abrasion. It arises in general when the abrasive particles are large or heavy, and contact the surface either with a certain amount of impact, or in a matrix which loads the particles against the surface. An example is the contact of hard flints in the earth against a digger blade. The appearance of gouging wear is shown in Fig. 6.1.

The stresses at the contact points are high, but the overall loading on the surface is low, and tends to be fairly uniformly distributed.

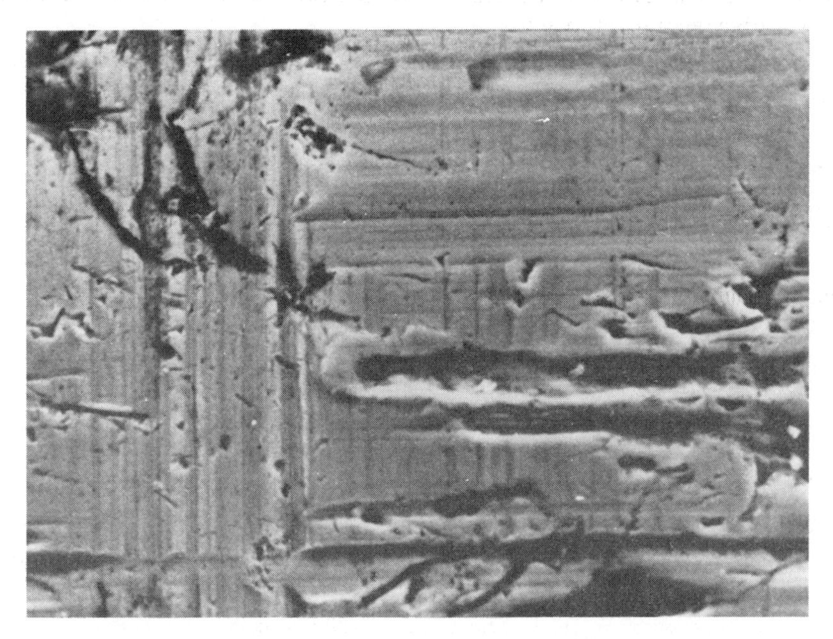

Grey Cast iron (×50)

FIG. 6.1 EXAMPLE OF GOUGING WEAR

TABLE 6.1 ALLOYS USED FOR RESISTANCE TO GOUGING ABRASION

Material	Hardness VPN	Properties	Applications
1. Tungsten carbide composites	~820	Maximum abrasion resistance: worn surfaces become rough: costly.	Rock drill bits: surface coatings: teeth of earth-handling equipment.
2. High-chromium irons (up to 32% Cr)	800–850	Excellent abrasion resistance: oxidation resistance.	Coal pulverising hammers: heat and corrosion resistant applications.
3. Martensitic iron	600–700	Excellent abrasion resistance: high-compressive strength.	Chute liners: roll crusher check plates: scoop lips: mill liners.
4. Cobalt base alloys	500–700	Oxidation resistance: corrosion resistance: hot strength and creep resistance.	Cutting tools, machine parts, roller burnishing: agricultural and earth-moving equipment, crushing equipment.
5. Martensitic low alloy steels	600–650	Good combination of abrasion and impact resistance.	Pulveriser hammers: earth-moving equipment.
6. Nickel base alloys	500–600	Corrosion resistance: may have oxidation and creep resistance.	Mill liners and chutes: impellers: scraper blades: pump plungers.
7. Pearlitic low alloy steels	300–450	Inexpensive: fair abrasion and impact resistance.	Impact mills, grates, screens, liners.
8. Stainless steels	up to 630	Corrosion resistance.	Shear blades, machinery parts: surgical and dental equipment, valve balls and seats, cutlery.
9. Austenitic steels: manganese steel	500 (by work-hardening)	Work-hardening: maximum toughness with fair abrasion resistance: good metal-to-metal wear-resistance under impact.	Weld repairs: hard facings: excavator buckets: mining, drilling, manufacturing and railway equipment: sprockets, pinions, gears, wheels and conveyor parts.

↑ Increasing wear-resistance

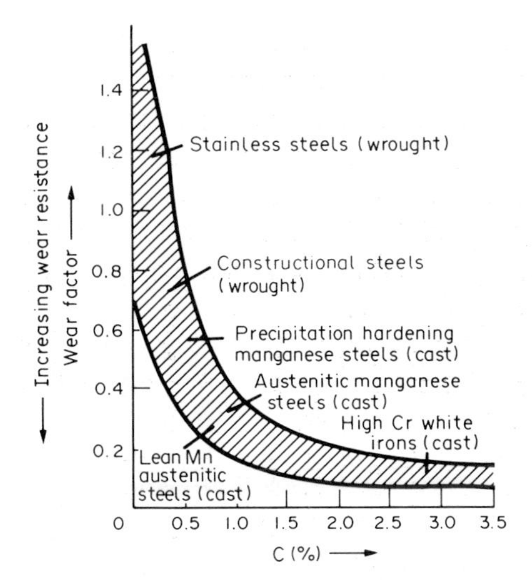

FIG. 6.2 EFFECT OF CARBON CONTENT ON THE RESISTANCE OF STEELS TO GOUGING WEAR

Significant lateral stresses are imposed on the surface, and because of the method of operation there may be intermittent impact forces. There is therefore a need for some toughness to resist fracture due to high impact or shear stresses, and this has to be balanced against the necessity for hardness to resist gouging.

Table 6.1 shows that, in general, the toughness rating of the available materials tends to be in reverse order to their abrasion resistance, and the choice for a particular application is a compromise between these two requirements.

6.2 MATERIAL SELECTION

A broad range of metallic materials resistant to gouging abrasion has been studied and the relationship between the wear ratio and carbon contents of various steels and irons undergoing gouging abrasion is shown in Fig. 6.2. The wear ratio is the ratio of the wear rate for the particular steel against that of a low alloy steel used as a standard. A general correlation exists between the carbon content and the resistance to gouging wear, a higher carbon content favouring a higher wear-resistance.

The effect of microstructure on gouging wear (Tables 6.3, 6.4) reflects the relative contributions of the matrix and the carbides. In general, steels and cast irons with martensitic matrices show greater resistance to gouging wear than those with ferritic or austenitic matrices. A study of the lean manganese austenitic steels showed that

TABLE 6.2 TUNGSTEN CARBIDE COMPOSITES

Type	% Composition WC	Co	TiC	Microhardness kg/mm^2	Typical applications
Tungsten carbide–cobalt	94	6	—	1500–1700	Metal forming tools, cutting tools, rotary drilling, percussion drills, chisels, hammers, cutting saws, machine construction, valve bodies, scrapers and nozzles.
	89	11	—	1300–1400	
	80	20	—	1050–1150	
	75	25	—	900–1000	
Tungsten carbide–titanium carbide	78	6	16	1600–1700	As above, but better performance where higher temperatures are encountered.
	78	8	14	1550–1650	
	86	9	5	1450–1550	

Notes:
1. Metal matrix can be either cobalt or nickel.
2. Increasing the % Co increases the toughness but lowers the hardness and wear-resistance.
3. Properties vary with carbide particle size—small particles ($<6\ \mu$) give high hardness with low toughness while coarser particles ($\sim 15\ \mu$) have lower hardness but higher toughness.
4. Addition of chromium to produce chromium carbides gives good corrosion resistance.

an increase in the volume of carbides also increases the wear-resistance.

In general, the hardness of work-hardened surfaces correlates well with the wear ratios, higher hardness values favouring higher wear-resistance.

Table 6.1 lists a number of alloys which have been developed commercially to resist gouging abrasion. The most wear-resistant materials are towards the top but toughness increases towards the bottom of the list. The hardest materials can be used as inserts or facings for tough steel backings, to compensate for their own lack of toughness.

6.2.1 Tungsten carbide composites (Table 6.2)

These materials have the highest hardness and best wear-resistance but are relatively costly. The cost factor, however, may be more than compensated for by the long life that can be obtained. Tungsten carbide is usually applied in the form of inserts in such applications as rock drill bits and facings on earth handling and digging equipment. The average deposit hardness of a tungsten carbide hardfacing can be up to 820 Vickers (65 RC).

Tungsten carbide is also used in a sintered form consisting of hexagon platelets fixed to a steel backing plate by soft or hard soldering. The platelets consist of tungsten carbide with a cobalt binder which is furnace sintered. The composition would be approximately 92% tungsten carbide, 6% cobalt and a residue of added carbides.

There appears to be no published information on the use of other ceramics to resist gouging wear.

6.2.2 High chromium irons (Table 6.3)

Irons containing up to 32% chromium and high carbon are available in many different alloy compositions (see Chapter 5). The three major groups, in order of increasing resistance to high stress abrasion are:

1. Austenitic types.
2. Hardenable grades.
3. Modifications containing tungsten, molybdenum or vanadium.

The austenitic types are relatively cheap and are best for metal-to-metal wear or low stress abrasion applications, e.g. farm equipment in sandy soil.

TABLE 6.3 HIGH CHROMIUM AND MARTENSITIC WHITE IRONS

Composition range %							VPN	
C	Si	Mn	Cr	Mo	Cu	Ni	(Heat treated)	Typical applications
1.20/3.50	0.30/1.50	0.40/1.50	8.0/32.0	0.50/3.0	1.0/3.0	–/6.0	630/770	Impact mills, waste comminuting plant, sinter grates, chute linings.

Notes:
1. Various combinations of hardness and toughness can be obtained by using selected compositions and heat treatments (air, oil or salt bath quenching).
2. Mo, Cu and Ni suppress pearlite formation and improve the response to heat treatment.
3. Further information is given in Chapter 5, Table 5.3.

TABLE 6.4 COMPARISON BETWEEN MARTENSITIC HIGH CHROMIUM IRON AND PEARLITIC WHITE IRON

Composition %	Martensitic high chromium iron	Pearlitic white iron
C	2.6	3.4
Si	1.5	0.5
Mn	1.1	0.5
Cr	14.3	1.0
Mo	3.0	—
Wear rate lbs/1000 tons	0.22	0.43

The hardenable grades are tougher than the austenitic types and have excellent resistance to both low and high stress abrasion when heat treated. A comparison between a martensitic high chromium iron and a pearlitic white iron in gouging type of service is given in Table 6.4.

The 14% chromium iron has about twice the wear-resistance of the pearlitic white iron. This performance can probably be improved by increasing the carbon content.

The addition of tungsten, molybdenum and vanadium in the last group helps increase hot hardness and adds to abrasion resistance.

6.2.3 Martensitic irons (Table 6.3)

These are available in two major types, chromium–molybdenum and nickel–chromium. The combination of martensite and a carbide matrix provides a hard composite structure with good abrasion resistance, and published data indicate that martensitic iron is from two to four times as resistant as the unalloyed and low chromium white irons under conditions of gouging abrasion.

6.2.4 Cobalt-base alloys (Table 6.5)

In general these alloys are used where wear and abrasion resistance combined with resistance to heat and oxidation and/or corrosion is required.

The best abrasion resistance is provided by an alloy containing 54–55% cobalt, 30% chromium, 12% tungsten and 2.5% carbon which is very hard (550/600 VPN), has a low coefficient of friction and can be used up to 1080°C. This alloy is, however, not recommended for parts subject to high impact and typical applications are ensilage knives and coke pusher shoes.

TABLE 6.5 COBALT BASE ALLOYS

	% Composition							
Grade	C	W	Cr	Mn	Si	Co	VPN	Applications
Tough	1.00	4.00	30.00	0.50	0.50	64.00	400	Cutting tools, burnishing rollers, anti-friction bearings, wear strips.
Medium	1.50	9.00	30.00	0.25	0.75	58.50	485	
Hard	2.50	18.00	32.00	0.60	0.40	46.50	750	

Notes:
1. Hardness increases with increasing % C.
2. Can also be used as hard facing material.
3. Excellent hardness and strength at elevated temperatures.

Another cobalt alloy containing 28% chromium 4% tungsten and 1% carbon is recommended where abrasion is accompanied by thermal shock or impact, and where cutting edges are desired. Although it is the softer of the cobalt alloys it has very good abrasion resistance and is used in applications where resistance to heat and impact are required.

6.2.5 Nickel base alloys (Table 6.6)

These alloys are also used where abrasion resistance plus resistance to heat and/or corrosion is required. Different alloy types encompassing some 15 compositions are available for different products including gear teeth, scraper blades and pump plungers.

6.2.6 Martensitic steels (Table 6.7)

Martensitic steels have an ideal combination of low cost, hardness, strength, abrasion resistance, good impact resistance and relatively high toughness. These hardenable alloys have moderate resistance to low and high stress abrasion, but increasing the carbon and chromium increases the abrasion resistance and impact strength. Depending upon composition martensitic steels have been used in a number of applications such as sprockets, crusher rolls, pulveriser hammers, cutting tools, forming dies and earth moving equipment.

6.2.7 Pearlitic steels (Table 6.8)

These low alloy steels have been used over a period of time for abrasion resistant parts. Little standardisation exists in these steels, as indicated by the range of composition, the use of alloying elements

TABLE 6.6 TYPICAL NICKEL BASE ALLOYS

	Major constituents %								
Type	Ni	Cr	Mo	Cu	Si	Fe	Condition	VPN	Applications
1. Illium B	50.0	28.0	8.5	5.5	2.50/6.30	—	Aged	410–600	Corrosion and erosion resistant cutting blades, pump impellers, rotary and thrust bearings.
2. "S" Monel	60.0 min.	—	—	27.0/31.0	3.5/4.5	2.5 max.	Aged	320–400	Valve seats, liners, pump rods.
3. "H" Monel	61.0/68.0	—	—	27.0/33.0	2.7/3.7	2.5 max.	As cast	250–305	Bushings, nozzles, machinery parts.

Notes:
1. (a) Ageing temperature for maximum hardness varies with % Si (760/600°C—4.2/6.2% Si).
   (b) Structure is two phase complex silicon–boron precipitation from austenitic solid solution.
2. Harden by air or furnace cooling from 600°C.
3. Used for anti-seizing parts.

TABLE 6.7 MARTENSITIC LOW ALLOY STEELS

% Composition										Yield strength psi	hardness VPN	Applications
C	Si	Mn	S	P	Cr	Ni	Mo	V	B			
0.15/0.35	0.20/1.80	0.60/1.80	0.045 max.	0.045 max.	Nil/1.30	Nil/2.00	0.15/0.60	Nil/0.15	Nil/0.006	100–200 × 10^3	315–530	Low section thickness in conditions requiring combination of strength, ductility and heat resistance.

Notes:
1. Sufficient alloy content is required to allow thorough hardening after austenitising and quenching.
2. Steels usually water quenched and tempered to produce fully martensitic structure.
3. Oil or salt bath treatment may also be used.
4. Development of desired toughness depends upon care taken during steel making and deoxidation (non-metallic inclusions must be suitably shaped and distributed: presence of high S, P, H_2 and N_2 is particularly harmful).

TABLE 6.8 PEARLITIC STEELS

Composition range %							VPN	
C	Si	Mn	Cr	Mo	Ni	V	(Heat treated)	Typical applications
0.5/1.2	0.3/0.6	0.6/1.0	0.5/2.5	0.2/0.5	0.5/1.5	0.05/0.10	300–450	Grates, screens, liners.

Notes:
1. Supplied in the as-cast, air-hardened, oil quenched or chill-cast conditions.
2. Quenching and tempering produces martensitic structures of excellent hardness and toughness (90–120 ft lb).
3. Lower carbon variety gives the best combination of strength and toughness.

depending upon the individual preference of the producers. Steel with the carbon content on the lower limit is generally used for applications requiring strength and toughness, while the high carbon variety is mainly used for applications requiring increased impact resistance.

These steels can be supplied in the as-cast, air-hardened, oil quenched or chill-cast conditions, to give a range of combinations of toughness and hardness according to the service requirements.

The microstructures are essentially pearlitic, the high carbon grades containing free cementite and the chill-cast being bainitic.

6.2.8 Stainless steels (Table 6.9)

Of the various types of stainless steels, the hardest and most wear-resistant are usually the martensitic type. Depending upon the grade and heat treatment used, they can be produced in hardnesses ranging from 155 to 630 VPN, as compared to 165 VPN for austenitic and 170 VPN for the ferritic stainless alloys.

Martensitic stainless steels can be hardened by conventional quenching and tempering, the properties varying with tempering temperature. Because of the combination of hardness, strength and corrosion resistance obtainable with such steels they find a use in a variety of applications such as shear blades, loom parts and plugs.

6.2.9 Austenitic manganese steel (Table 6.10)

This material is commonly used for impact and abrasion resistance. Its important properties are its high toughness with the ability to work-harden under impact, hardnesses of around 600 VPN being achieved under surface loading from an initial hardness of around 200 VPN. It lends itself well to weld repair or to the build up of hard facings. However, due to the work-hardening property, it is difficult to machine and is therefore usually only available as castings. The material is used widely as lip material for excavator buckets in hard rock digging; however, if the abrasive medium is very coarse the steel may be abraded in depth prior to the occurrence of work-hardening, thus giving a shortened life. Where conditions of bulk plastic flow are likely, an addition of 2% molybdenum is made which increases the yield strength considerably and provides an even better wear resistance.

TABLE 6.9 STAINLESS STEELS

Nominal composition %								
C	Mn (max.)	S (max.)	Si (max.)	Cr	Mo	Heat treatment	VPN	Applications
0.15 (min.)	1.00	0.03	1.00	12.0–14.0	—	Hardened and tempered 320°C	540	Textile machine parts.
0.60–0.75	1.00	0.03	1.00	16.0–18.0	0.75 (max.)	Hardened and tempered 320°C	560	Shear blades.
0.75–0.95	1.00	0.03	1.00	16.0–18.0	0.75 (max.)	Hardened and tempered 320°C	600	Surgical and dental equipment.
0.95–1.20	1.00	0.03	1.00	16.0–18.0	0.75 (max.)	Hardened and tempered 320°C	630	Valve and bearing balls, races, instruments.

Notes:
1. Strength and hardness increase with carbon content.
2. Quenched in air or oil from temperature of ~1000°C then tempered at intermediate temperatures to obtain desired properties.
3. Good combination of hardness, strength and corrosion resistance.

TABLE 6.10 AUSTENITIC MANAGANESE STEELS

Material	C	Mn	Si	Cr	Mo	Ni	Strength psi × 10^3	BHN VPN 30 kg	Heat treatment	Applications
	% Element									
A	1.05–1.35	11.0 min.	0.3–1.0	—	—	—				Teeth and buckets of heavy duty excavators, crusher jaws, liners and screens. Blades for scraper scoops, beater bars in hammer mills, feeder pans, excavator pans.
B-1	0.9–1.05	11.5–14.0	0.3–1.0	—	—	—				
B-2	1.05–1.2	11.5–14.0	0.3–1.0	—	—	—	45–55	190–210	Water quenched from 1050°C	
B-3	1.12–1.28	11.5–14.0	0.3–1.0	—	—	—				
B-4	1.2–1.35	11.5–14.0	0.3–1.0	—	—	—				
C	1.1–1.3	12.0–14.0	0.3–1.0	1.5–2.5	—	—	50–60	200–335	Water quenched from 1100°C	
D	1.05–1.15	12.0–14.5	0.5–0.1	—	—	3.5–5.0	40–55	160–190	Water quenched from 1050°C	Digger teeth, key bars, liners.
E-1	1.0–1.2	13.0–14.0	0.4–0.6	—	0.9–1.1	—	52–62	180–210	Water quenched from 1050°C	Heavy duty crusher parts.
E-2	1.0–1.2	13.0–14.0	0.4–0.6	—	1.8–2.2	—	60–70	180–210	12 hrs 600°C Water quenched from 1000°C	

Notes:
1. Excellent toughness in the non work-hardened condition giving good resistance to gouging abrasion.
2. Chromium increases hardness and yield strength.
3. Molybdenum considerably increases yield strength and stabilises the embrittling action of undissolved carbides.
4. E-2 material is costly, and application so far is limited.
5. Copper may be added to give improved strength, toughness and work hardening.
6. Titanium sometimes added as a grain refiner.

6.3 GUIDELINES

General guidelines to the selection of materials to resist gouging wear are given in Table 6.11, and guidelines to compositional and microstructural factors in Tables 6.12 and 6.13 respectively.

Table 6.14 gives guidelines to the effect of abrasive particle properties on the selection of materials to resist gouging wear.

TABLE 6.11 GENERAL GUIDELINES TO THE SELECTION OF MATERIALS TO RESIST GOUGING WEAR

1. For resistance to gouging abrasion there is a need for toughness and hardness. Since toughness often decreases as hardness increases, the choice will usually be a compromise between the two requirements.
2. The resistance of a technically pure metal to gouging wear is related to its hardness in the fully annealed condition.
3. For most practical metals the wear resistance is controlled by work-hardening and does not vary greatly with initial hardness.
4. Hardness is only effective in increasing gouging wear resistance if it is greater than about half that of the abrasive.
5. If the hardness of the surface is greater than about 1.3 times the hardness of the abrasive, little or no further gouging wear-resistance is obtained by further increase in hardness.
6. Surface coatings or hard facings can be used to provide improved gouging wear-resistance, but the depth must be related to the permissible wear depth (see Chapter 10).

TABLE 6.12 GUIDELINES TO COMPOSITIONAL FACTORS IN THE SELECTION OF MATERIALS TO RESIST GOUGING WEAR

1. Wear-resistance of carbon and low alloy steels increases with increasing carbon content.
2. Small amounts of manganese improve the wear resistance of pearlitic structures but reduce that of martensitic structures.
3. Steels with high manganese contents ($\sim$12%) have very good impact and wear-resistance. However, the conditions must be such that the work-hardening rate is greater than the wear rate so that the austenite to martensite change takes place.
4. The wear resistance of pearlitic steels generally increases with the addition of carbide-forming elements.
5. Vanadium and niobium carbides give the best resistance to penetration of alloy steel surfaces.

TABLE 6.13 GUIDELINES TO MICROSTRUCTURAL FACTORS IN THE SELECTION OF MATERIALS TO RESIST GOUGING WEAR

1. For a given carbon content, gouging wear decreases with increased grain size of a steel.
2. The best structure for gouging wear-resistance in carbon steels appears to be lower bainite.
3. In hypoeutectoid steels (<0.85% C) wear-resistance increases with increased percentage volume fraction of pearlite in the structure and with decrease in interlamellar spacing.
4. In hypereutectoid steels (>0.85% C) wear-resistance increases with carbon content provided that Fe_3C does not occur as a grain boundary network.
5. For alloy steels the form of the carbide is the most important factor. Removal of primary carbides and precipitation of fine uniform secondary carbide in the matrix gives high wear resistance.
6. For alloy steels the best wear-resistance is obtained with uniformly distributed fine carbide in a martensitic matrix with some residual austenite.
7. With carbon steels isothermal heat treatment gives toughness and wear-resistance.
8. Martempering gives high wear-resistance but leaves residual microstresses.

TABLE 6.14 GUIDELINES TO THE EFFECT OF ABRASIVE PARTICLE PROPERTIES ON THE SELECTION OF MATERIALS TO RESIST GOUGING WEAR

1. If the abrasive particles are fine, the presence of coarse hard particles in alloy can give considerable improvement in wear-resistance.
2. If the abrasive is coarse a more homogeneous alloy structure is desirable.
3. If the abrasive is fine a brittle metal or ceramic can be used, but with coarse abrasives surface cracking can lead to increased wear.

Chapter 7

Low-stress Abrasion (Category 5)

7.1 INTRODUCTION

Low-stress abrasion manifests itself in solids-handling equipment such as conveyors, chutes, vibrating screens, etc. Since impact is virtually absent the property of toughness as associated with gouging abrasion is not so important, the important properties being hardness, together with corrosion resistance, cost and ease of replacement.

An example is shown in Fig. 7.1.

A wide range of materials (cast irons, cast steels, rolled steel plate, hardfacings, ceramics, concretes, rubbers and plastics) have been used in order to combat wear conditions of low stress abrasion by sliding particles.

Silica on Steel (×50)

FIG. 7.1 EXAMPLE OF LOW-STRESS ABRASION

TABLE 7.1 RELATIVE RESISTANCE OF VARIOUS MATERIALS TO LOW-STRESS ABRASION BY COKE (LOW DENSITY)

Material	Wear index
Sintered tungsten carbide	0.028
Fusion-cast alumina, special	0.090
Fusion-cast alumina	0.115
High-chrome hardfacing	0.207
High-chrome martensitic white cast iron No. 1	0.231
Ni–Cr martensitic white cast iron	0.240
Cr–Mn–Mo hardfacing	0.243
13% Mn skin-hardened steel plate	0.307
Slagceram	0.326
High-chrome cast iron No. 2	0.328
30% Cr–Mo cast steel	0.370
Cast basalt	0.378
Acid-resisting ceramic tile	0.413
Low-alloy steel plate, quenched and tempered, No. 1	0.460
Nodular graphite-based cast iron	0.470
13% Mn austenitic cast steel	0.476
Low-alloy steel plate, quenched and tempered, No. 2	0.525
Low-alloy cast iron	0.535
Low-alloy steel plate, quenched and tempered, No. 3	0.665
High-phosphorus pig iron	0.700
Concrete	1.900–0.910
En8 steel plate	1.000
Solid rubbers	4.800–1.850

Table 7.1 provides an example of wear indices of this range of wear-resistant materials when subjected to sliding wear by an abrasive substance, in this case coke. The wear index is the ratio of the wear rate of the material to the wear rate of a standard material, En8 steel plate.

The results show that sintered tungsten carbide is by far the most wear-resistant material when subjected to sliding wear by coke. It suffers, however, from the disadvantage of being very costly and the fusion-cast alumina special becomes the best material with regard to cost/life ratio. Using the fusion-cast alumina special instead of a standard En8 steel, against sliding abrasion by coke, the life of the wear surface can be increased by a factor of more than ten. The results also show that the ability to work-harden is beneficial, work-hardened 13% Mn steel being about 50% more wear-resistant than the non-work-hardened counterpart.

TABLE 7.2 RELATIVE RESISTANCE OF VARIOUS MATERIALS TO LOW-STRESS ABRASION BY SINTER (HIGH DENSITY)

Material	Wear index
Ni–Cr martensitic white cast iron	0.065
Cr–Mo–W–Nb hardfacing	0.079
High-chrome martensitic white cast iron No. 1	0.115
Nodular graphite-based cast iron	0.137
Fusion-cast alumina special	0.172
Sintered alumina	0.173
Tungsten carbide hardfacing	0.192
Fusion-cast alumina	0.221
Low-alloy steel plate, quenched and tempered No. 2	0.382
Slagceram	0.518
Low-alloy cast iron	0.704
Fusion-cast basalt	0.846
Low-alloy steel plate, quenched and tempered No. 1	0.952
High-phosphorus pig iron	0.953
En8 steel plate	1.000
Low-alloy steel plate, quenched and tempered No. 3	1.336
Acid-resisting ceramic tile	2.030
Silicon carbide ceramic No. 1	2.265
Rubber	2.490
Silicon carbide ceramic No. 2	4.710
Silicon carbide ceramic No. 3	8.470
Concrete	18.000

The wear index and the basic material costs are not the only criteria when considering the selection of wear-resistant materials. Other factors such as availability, ease of replacement and time required for repair must also be taken into account. Thus when considering the results in Table 7.1, although materials such as concrete and rubber appear in the lower levels, they are useful as wear-resistant materials if lowest material cost is of prime importance. Rubber linings are found to be effective in screw conveyors providing the screws are coated with, or manufactured from, hard materials.

Table 7.2 gives the wear indices of similar groups of materials when subjected to sliding wear by abrasive particles of higher density, in this case blast furnace sinter, which is approximately four times as dense as coke.

The ceramic materials which behaved well under sliding abrasion by coke did not perform so well when subjected to sliding wear by sinter. This is probably due to the higher density of the sinter, which would affect the wear mechanism prevailing, possibly introducing some need for toughness.

Thus the choice of the best material to resist low-stress abrasion depends to some extent on the properties of the abrasive, and there is not enough published information to enable this effect to be predicted. It appears, however, that in general white cast irons will be among the best materials.

For steels and cast irons in general, the harder the material the more likely it is to have a long life under low-stress abrasion conditions. The longest-lasting wear resistant materials are generally the most economical in the long run, since they bring the added benefits of reduced labour and downtime costs.

In situations where high temperatures are involved, materials with high chromium content are worth considering. Hardfacing materials are also potentially useful for applications where only a comparatively small quantity is required, e.g. for fan blades.

7.2 GUIDELINES

General guidelines to the selection of materials to resist low-stress abrasion are given in Table 7.3, and guidelines to compositional and microstructural factors in Tables 7.4 and 7.5.

Table 7.6 gives guidance to the effect of abrasive particle properties on the selection of materials to resist low-stress abrasion.

TABLE 7.3 GENERAL GUIDELINES TO THE SELECTION OF MATERIALS TO RESIST LOW-STRESS ABRASION

1. For resistance to low-stress abrasion the dominant requirement is for hardness.
2. The resistance of a technically pure metal to low stress abrasion is related to its hardness in the fully annealed condition.
3. For most practical metals the wear resistance is controlled by work-hardening and does not vary greatly with initial hardness.
4. Hardness is only effective in increasing abrasion resistance if the hardness is greater than about half that of the abrasive.
5. If the hardness of the surface is greater than about 1.3 times the hardness of the abrasive, little or no further abrasion resistance is obtained by further increase in hardness.
6. Work-hardening is a useful characteristic because the wear is likely to be slow enough for work-hardening to develop fully.
7. Ceramics are the most resistant materials for low-density abrasives, but are less satisfactory for high-density abrasives, where there may be a tendency for surface cracking.
8. Hard white cast irons are the most resistant materials for high-density abrasives.
9. Nodular graphite cast iron is effective with dense abrasives.

TABLE 7.4 GUIDELINES TO COMPOSITIONAL FACTORS IN THE SELECTION OF MATERIALS TO RESIST LOW-STRESS ABRASION

1. Wear-resistance of carbon and low alloy steels increases with increasing carbon content.
2. Small amounts of manganese improve the abrasion resistance of pearlitic structures but reduce that of martensitic structures.
3. Steels with high manganese contents (~12%) have very good impact and abrasive wear-resistance. However, the conditions must be such that the work-hardening rate is greater than the wear rate so that the austenite to martensite change takes place.
4. The abrasion resistance of pearlitic steels generally increases with the addition of carbide-forming elements.
5. Vanadium and niobium carbides give the best resistance to abrasive penetration of alloy steel surfaces.

TABLE 7.5 GUIDELINES TO MICROSTRUCTURAL FACTORS IN THE SELECTION OF MATERIALS TO RESIST LOW-STRESS ABRASION

1. For a steel of a given carbon content increasing the grain size will decrease wear by low-stress abrasion.
2. Lower bainitic structures in carbon steels appear to be best for resisting wear by low-stress abrasion.
3. The wear-resistance of hypoeutectoid steels (<0.85% C) increases with increased percentage volume fraction of pearlite in the structure and with a decrease in the interlamellar spacing.
4. The wear-resistance of hypereutectoid steels (>0.85% C) increases with carbon content provided that the Fe_3C does not occur as a grain boundary network.
5. The wear resistance of alloy steels is greatly influenced by the form of the carbides present. The removal of the primary carbides and the precipitation of fine uniform secondary carbides in the matrix will produce a high wear-resistance.
6. Resistance to low-stress abrasion is best in alloy steels when the structure consists of a uniformly distributed fine carbide in a martensitic matrix containing some residual austenite.

TABLE 7.6 GUIDELINES TO THE EFFECT OF ABRASIVE PARTICLE PROPERTIES ON THE SELECTION OF MATERIALS TO RESIST LOW-STRESS ABRASION

1. If the abrasive particles are fine, the presence of coarse hard particles in alloy can give considerable improvement in abrasion resistance.
2. If the abrasive is coarse, a more homogeneous alloy structure is desirable.
3. If the abrasive is fine, a brittle metal or ceramic can be used, but with coarse abrasives surface cracking can lead to increased wear.

Chapter 8

Erosion (Category 6)

8.1 INTRODUCTION

Erosion is the type of wear damage which is produced by the impingement of sharp particles on a surface. The particles are often transported in a moving liquid or gas, and examples are the transport of slurries in pipelines, movement of pump impellers in slurries, shot-blasting, or exposure to sandstorms. The presence of moisture in the conveying fluid also poses the additional problem of corrosion, and in many cases this can be a stronger failure mechanism than the erosion itself. An example of erosion is shown in Fig. 8.1.

Two damage mechanisms contribute to erosion. The first is abrasive erosion which is the micromachining action of abrasive particles

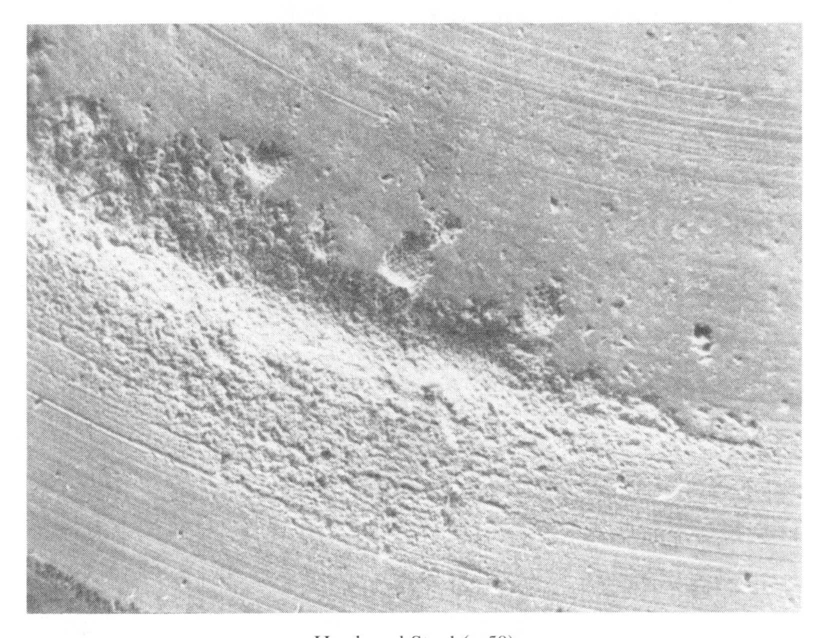

Hardened Steel (×50)

FIG. 8.1 EXAMPLE OF EROSIVE WEAR

which impact the solid surface at a small angle, i.e. relative motion of the abrasive particles is nearly parallel to the solid surface. This is a low-stress abrasive process where impact does not generate sufficient stress to fracture the abrasive. It is also commonly termed "scratching abrasion". The second is impingement erosion where impact of the abrasive particles occurs at a large incident angle.

When considering materials to withstand erosion the properties of the abrasive particles, namely velocity, impact angle, quantity, density, hardness and surface contour, are important. As the stresses involved with erosive wear are only occasionally great enough to break the abrasive particles it follows that the original smoothness or angularity is important, since there is little change with use.

8.2 EFFECT OF IMPACT ANGLE

The influence of impact angle is dependent upon the type of material undergoing erosion. For ductile materials the maximum damage occurs at low impact angles whereas for brittle materials it occurs at normal impact. Figures 8.2 and 8.3 illustrate the influence of the impact angle on various materials used to combat erosion.

8.2.1 Low impact angles

With low impact angles, i.e. in abrasive erosion, the best choice for a wear-resistant material is to use as hard a material as possible, e.g. hardened steel of high carbon content, cast irons or ceramics. With steels a high carbon content ensures a high proportion of hard carbides within the structure. Figure 8.4 is a general curve of hardness versus carbon content and it can be seen that hardness increases rapidly up to a carbon content of 0.7%. Above this value only a small hardness increase is gained for a large increase in carbon.

White cast irons are generally the choice for resistance to scratching or abrasive erosion. Pearlitic white and chilled irons have been, and still are being, used to a considerable extent although they are generally inferior to higher alloy irons. The higher wear-resistance of martensitic white irons over the pearlitic irons in most applications means that the former are now tending to replace the latter in erosive situations and are proving more economical on a service life basis. Table 8.1 lists various hard materials in common use to resist abrasive erosion.

Thus for systems involving small impact angles, where cutting wear prevails, the use of hard materials is sufficient, e.g. straight runs in slurry pipelines. When using hard materials the erosion of bends in

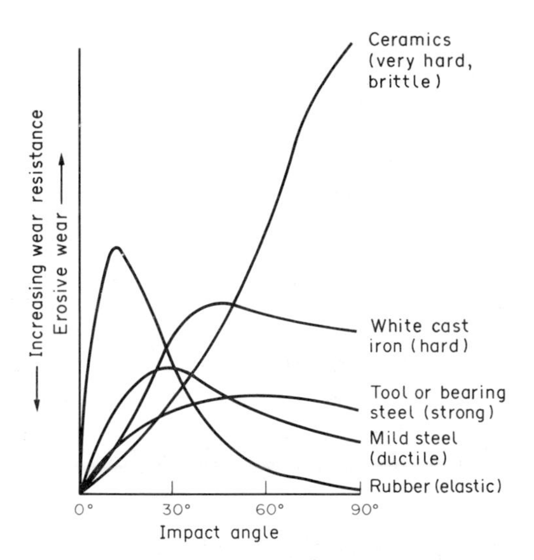

FIG. 8.2 EFFECT OF IMPACT ANGLE ON EROSIVE WEAR

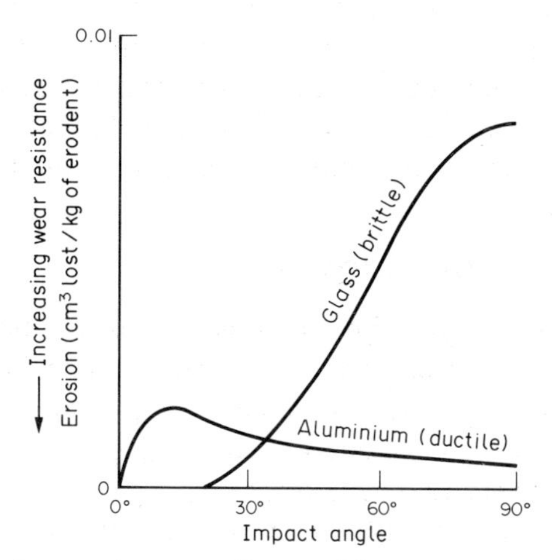

FIG. 8.3 EFFECT OF IMPACT ANGLE ON EROSIVE WEAR

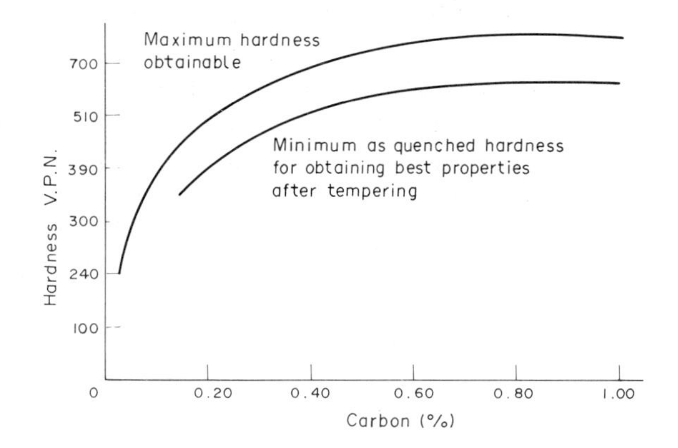

FIG. 8.4 HARDNESS AS A FUNCTION OF CARBON CONTENT

TABLE 8.1 TYPICAL MATERIALS FOR EROSION RESISTANCE

Particle properties	Typical materials	Reference
Low speed, low angle	High chromium white cast irons	Table 5.3
	Martensitic alloyed cast irons	Table 5.4
	Martensitic steels	Table 5.2
	Ceramics	Table 5.11
	Plastics	Table 8.2
High speed, low angle	Nickel base alloys	Table 6.6
	High carbon chrom-moly steel	Table 5.2
	Ceramics	Table 5.11
High speed, high angle	11% chromium steel (0.16 C, 11 Cr, 6 Mo, 0.25 Nb, 0.3 V)	
	Nickel alloys (20 Cr, 20 Co, 2.5 Ti, 5 Fe, 1.5 Al, 1.5 Si, 1.0 Mn)	
	Hot pressed silicon nitride (s.g. 3.1–3.16)	
Low speed, high angle	Rubbers and resins	Table 8.2

transport lines can be reduced by using large radii of curvature so that the product of particle velocity and sine of the impingement angle does not exceed the maximum particle velocity (K) at which the collision remains purely elastic.

8.2.2 High impact angle

When the impact angle is large, i.e. in impingement erosion, wear due to repeated deformation takes place. Prevention of erosion in this situation is not so straight-forward. Hard materials may be used at low particle velocities. For conventional materials of construction the impact velocity values for which the collision remains purely elastic rarely exceeds 10 m/sec. For soft rubber however, the limiting impact velocity is much larger due to its low elastic modulus and this coupled with a large deformation wear factor (ε) makes it a suitable material for use in many impingement erosion applications, e.g. for backing sheets for shot blasting booths. Table 8.2 lists some elastic materials to resist impact erosion.

8.3 EFFECT OF IMPACT VELOCITY

For a given situation the weight loss by erosion increases as the particle velocity is increased. Figure 8.5 illustrates the effect of increasing the impact velocity on the volumetric wear rate of various materials.

TABLE 8.2 ELASTIC MATERIALS FOR IMPACT EROSION RESISTANCE

Material	Formulation	Incompatible with	Applications
Natural rubber, SBR (Buna S)	With carbon black, high cross-linking.	Strong acids, solvent, mineral oil halogens.	Pipelines for aqueous slurries, shot blast cabinets, poor resistance to ozone.
Nitrile rubber	High nitrile, high cross-linking.	Ozone, oxygen-rich atmospheres, oxidising acids.	Slurries or pumps where mineral oils present, cyclone driers to 150°C, shot blasting cabinets.
PVC	Plasticised.	Organic solvents, strong acids.	Pipelines for aqueous slurries, pump casings.
Nylon, nylon 66	Unreinforced.	Strong acids, halogenated solvents.	Impeller blades, fan blades, agitators.
Neoprene rubber	Highly cross-linked.		Low impact speed (e.g. dust extractors).
Fluorocarbon rubber	Cross-linked.		Heat, chemical or oil environments.
Resins, epoxy phenolic ABS	Highly cured, low filler content, enough reinforcement for required structural strength.	Strong acids, halogens.	Pipelines, impellers, fan blades, pump casings. Used as coatings, and especially for rebuilding eroded sections.

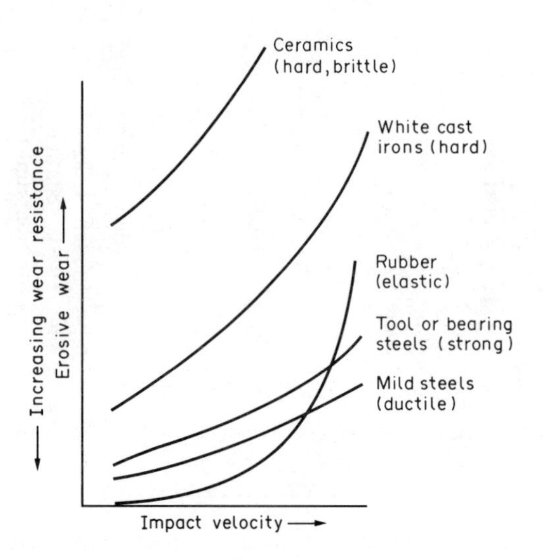

FIG. 8.5 EFFECT ON IMPACT VELOCITY ON EROSIVE WEAR

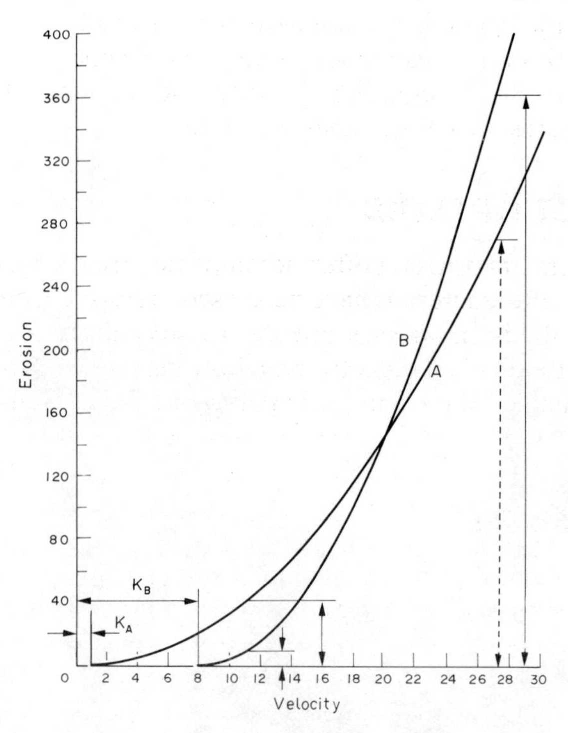

FIG. 8.6 INFLUENCE OF VELOCITY ON DEFORMATION WEAR

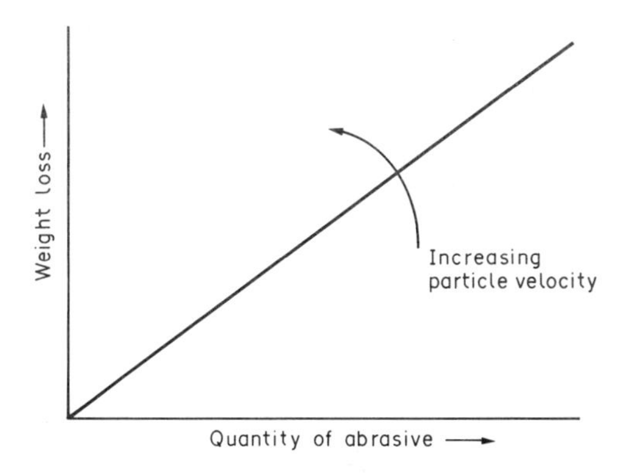

FIG. 8.7 EFFECT OF QUANTITY OF ABRASIVE AND PARTICLE VELOCITY ON WEIGHT LOSS IN EROSIVE WEAR

In impingement or deformation wear the effect of increasing particle velocity depends upon the type of material being deformed, i.e. whether a soft ductile material (small K, high ε) is being eroded. From Fig. 8.6 it is seen that at low particle velocities the hard substance B shows less erosion since K has a large influence. At higher velocities the reverse is true since K is now negligible and ε has a larger influence.

Inceasing the quantity of abrasive will increase the weight of material lost by erosion, all other parameters remaining constant, as shown in Fig. 8.7. Increasing quantity and velocity of abrasive particles gives rise to a more rapid wear rate.

8.4 EFFECT OF FLUID

The problem of corrosion often accompanies erosive systems, and a number of softer materials have been used to resist chemical attack by different fluids. Examples are the use of rubber as linings or fabrication of individual units from high strength plastics in aqueous systems. Examples of the application of the softer materials are given in Table 8.2.

8.5 SUMMARY

Thus to combat erosion, two completely different approaches can be used, either to use hard materials which are resistant to abrasive erosion, or to use soft materials which will absorb the energy of impact in impingement erosion. These approaches can be used separately or in combination depending on the erosive situation facing the designer, as shown in Fig. 8.8.

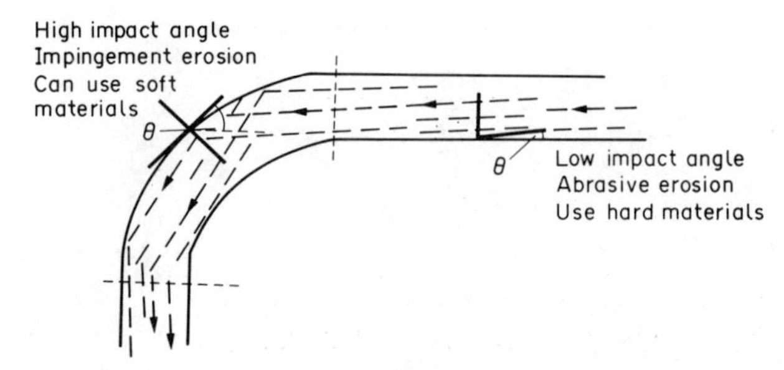

FIG. 8.8 SELECTION OF MATERIALS FOR PIPELINES OR CHUTES, SUBJECT TO EROSIVE WEAR

8.6 GENERAL GUIDELINES

General guidelines for the selection of materials to resist particulate erosion are presented in Table 8.3. The relationship between impact angle and velocity and the required material properties is shown graphically in Fig. 8.9.

TABLE 8.3 GENERAL GUIDELINES FOR THE SELECTION OF MATERIALS FOR RESISTANCE TO EROSION

1. The selection of materials to resist erosion depends on the angle at which the particles strike the surface, the impact velocity, the nature of the particles, and the nature of the fluid in which the particles move.
2. At low impact angle and low-speed the main selection criterion is hardness of the material (see Fig. 6.5).
3. At low impact angle and high speed the requirement is for hardness with adequate toughness (see Fig. 6.5).
4. At high impact angle and low speed the main requirement is for elasticity (see Fig. 6.5).
5. At high impact angle and intermediate speed there is an increasing need for resilience to resist penetration of the surface by the particles (see Fig. 6.5).
6. At high impact angle and high speed there must be strength with ductility to permit repeated deformation without loss of material (see Fig. 6.5).
7. Harder particles require harder surfaces.
8. The surface must not be attacked physically or chemically by the transporting fluid (see Table 6.4 for compatibility of non-metallic materials).
9. Where rubber is used it must be tough enough to resist penetration by the erosive particles, but not so hard that it is no longer elastomeric. Recommended hardness 50–90 degrees.
10. Rubber resilience should be high, so hardness should be achieved by cross-linking rather than by fillers.

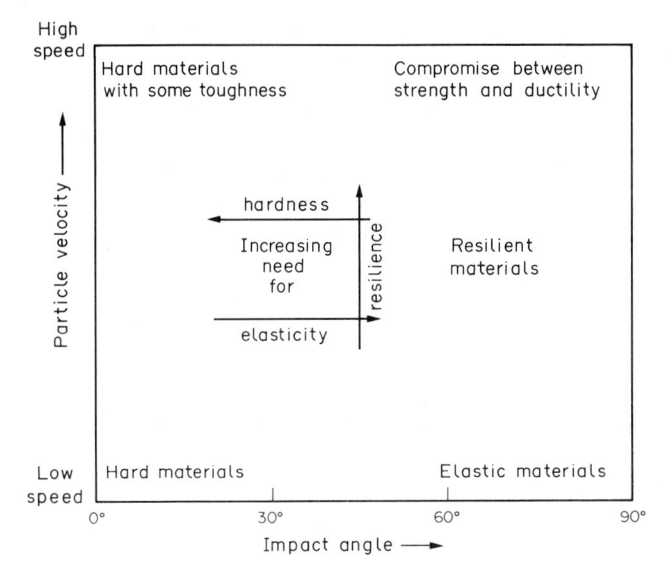

Note: See Table 8.1 for typical materials

FIG. 8.9 GUIDE TO SELECTION OF MATERIAL TYPE ACCORDING TO PARTICLE SPEED AND IMPACT ANGLE

Chapter 9

Corrosive Wear (Category 7)

9.1 INTRODUCTION

Because of the complexity of corrosive wear in general, and the small amount of co-ordinated work which has been done on the subject, it is not possible to provide simple guidance to the selection of materials to resist corrosive wear. This chapter is intended only to provide some brief guidance, and references to sources of further information.

9.2 NATURE OF CORROSIVE WEAR

Corrosive wear has been defined as a wear process in which chemical or electrochemical reaction with the environment predominates. This is not a satisfactory definition because the corrosive and wear functions may be intimately linked together, as shown schematically in Fig. 9.1. One or both of the types of attack on the original surface may be very low, and one or other may be missing completely, and the overall process is likely to be dominated by corrosion of worn surfaces and wear of corroded surfaces, neither of which can proceed faster than the other.

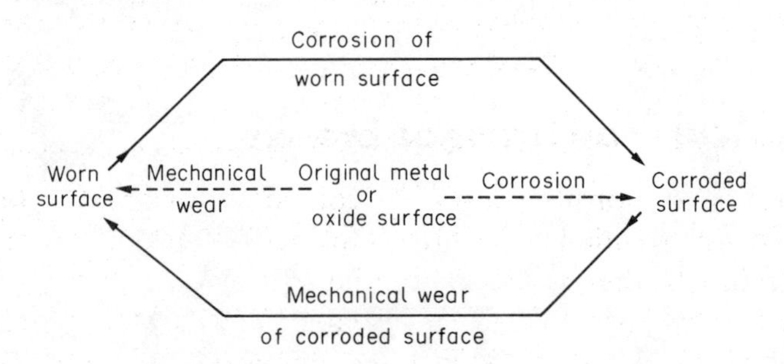

FIG. 9.1 GENERAL MECHANISM OF CORROSIVE WEAR

Because the wear rate of a corroded surface will often be higher than that of an uncorroded surface, and the corrosion of a worn surface will often be higher than that of an unworn surface, the total rate of loss of material in corrosive wear can be high and the resulting problems can be very serious.

Technically, any definition of corrosive wear includes the re-oxidation of exposed metal in a worn surface, and in general this is a beneficial phenomenon. The same is true of the action of extreme pressure additives in lubricants. For the purposes of this guide corrosive wear will be taken only to include those situations where the chemical attack leads to an undesirable result rather than a desirable one.

All chemical reactions take place more rapidly at higher temperatures, and a rough rule of thumb is that the reaction rate doubles for a 10°C rise in temperature. It follows that there is a general tendency for corrosive wear problems to be more serious at high temperatures.

9.3 IDENTIFYING THE NATURE AND CAUSE OF CORROSIVE WEAR

The wear debris in corrosive wear tends to be finely divided and fully reacted, with very little or no unreacted metal present. The existence of such debris is not a definite indication of corrosive wear, because in abrasive wear it is possible for the wear debris to be fully transformed before or after removal from the surface.

The worn surface will be fairly smooth, but the total loss of material can be very high. There will often be traces of corrosion on the worn surface, and it is important for the surface to be examined as soon as possible, before atmospheric corrosion can confuse the issue.

It will sometimes be obvious that a corrosive situation exists, but this is not always the case. The corrosive material will sometimes only attack the surface once wear has taken place, and will therefore not immediately be recognised as corrosive. There are an almost infinite number of possible situations, but the following three examples will illustrate the sort of problem which can arise.

9.3.1 Water on brass or bronze

Water is not normally a serious corrodent for brass or bronze, as the oxide/hydroxide layer which is formed will provide some protection against further attack. If the layer is subject to wear it can be removed very easily because it is friable. The surface is then open to further attack by water, which under conditions of wear is therefore a corrodent for brass or bronze.

9.3.2 Dilute hydrochloric acid on steel

Under normal circumstances very dilute solutions of hydrochloric acid in water are not corrosive to steel, because the initial attack produces a passive film on the surface. Under wear conditions, however, the passive film can be worn away, and the surface exposed to fresh attack.

9.3.3 Oxidative wear of stainless steels

Stainless steels derive their oxidation resistance from the formation of a barrier layer of oxide of the alloying metal (e.g. chromic oxide). However, the formation of this oxide layer depends on a relatively slow diffusion process, and under rubbing conditions the layer may not have time to form before it is worn away. Thus under oxidative wear conditions stainless steels may be little more resistant than ordinary carbon steels.

It may therefore be necessary to show experimentally that corrosive wear is taking place. This can be done by performing a laboratory wear test with and without the various components of the service environment, until the corrosive component can be identified. Alternatively it can be done by changing some component of the wear situation in service.

In general corrosive wear should be considered wherever excessive wear is taking place and the debris is fully reacted, especially at high temperatures or where substances are known to be present which react even mildly with the material which is being worn.

Table 9.1 lists some of the more common corrodents which contribute to corrosive wear, but the range of possible corrodents is enormous.

TABLE 9.1 TABLE OF COMMON CORRODENTS IN CORROSIVE WEAR

1. Water, at temperatures between 0°C and 150°C.
2. Oxygen, especially above 150°C.
3. Atmospheric hydrogen sulphide.
4. Atmospheric sulphur dioxide.
5. Sea water.
6. Mineral acids.
7. Organic acids.
8. Many process liquids or gases.

9.4 PREVENTING CORROSIVE WEAR

From Fig. 9.1 it can be seen that any action which interrupts the corrosion/wear cycle will prevent corrosive wear. In practice it is usually found that the most effective action is to prevent the corrosion rather than the wear. To achieve this effectively requires an analysis of the corrosion taking place, and it is not possible in this guide to cover all the possible situations. Table 9.2 lists some of the actions which should be attempted, but it must be recognised that these will not be easy or even possible in all cases.

TABLE 9.2 POSSIBLE ACTIONS TO REDUCE CORROSIVE WEAR

1. Identify nature of corrosive reaction.
2. Eliminate corrosive substances (e.g. driers to remove water, partial inerting to reduce oxygen).
3. Reduce temperature.
4. Change surface material:
   (a) If wear process is mild abrasion, try paint or epoxy primer.
   (b) If wear process is erosive, try elastomer (rubber) instead of metal.
   (c) Try changing from steel to copper alloy, or vice versa.
   (d) If hard surface is required, try ceramic.
   (e) Change from metallic to non-metallic material, or vice versa.
5. Try to reduce the wear problem by reducing speed, load or presence of abrasives.

Chapter 10

Wear-resistant Surfaces and Coatings

10.1 INTRODUCTION

Numerous situations exist where improved wear-resistance is required in components in order to maintain specific dimensions or give improved life, and here the optimum solution is to provide a hard wear-resisting surface on a cheaper or tougher core. Table 10.1 illustrates the variety of the major treatments available, to what materials they are applicable and the chief characteristics of each treatment.

10.2 DIFFUSION COATINGS

Diffusion coatings are produced by bringing suitable elements in contact with the surface of a metal at high temperatures so that those elements diffuse into the surface, change its composition and improve its properties. Diffusion treatments are widely used to improve wear and abrasion resistance as well as corrosion and heat resistance. By incorporating various elements into the base metal it is possible to combine a hard and wear-resistant surface with a low cost, strong and tough base. Further heat treatment in many cases can further improve core characteristics.

10.2.1 Carburising

The carbon content of the surface layer can be increased by means of a solid, liquid or gas carrier. The process is applied to steels with a low initial carbon content (<0.45%), which readily absorb more carbon.

Depending upon the metal composition and processing variables, the carburised surface can have a hardness of up to 800 VPN and a case depth from less than 0.5 mm (0.020 in) to more than 1.5 mm (0.060 in).

TABLE 10.1 MAJOR HARD COATINGS AND SURFACES

Type of coating or surface	Metals that can be coated	Major characteristics
Diffusion coatings		
Carburized	Carbon and alloy steels with low carbon content.	High core strength and toughness combined with extreme surface hardness.
Nitrided	Nitriding steels, some medium carbon steels, stabilised stainless steels, high speed tool steels, some die steels and cast irons.	High wear-resistance, retention of hardness at elevated temperatures.
Cyanided and carbonitrided	Many of same metals as carburising.	Generally same as carburised cases.
Chromised	Carbon steels, many alloy steels, ductile iron, powder metals, A2 tool steel, stainless steel.	High hardness and resistance to wear.
Siliconised	Low carbon, low sulphur steels.	High wear-resistance: some swelling.
Boronising	Steels, cast irons, refractory metals.	Excellent sliding wear-resistance.
Electroplates		
Hard chromium	Most ferrous and non-ferrous metals.	High hardness, low coefficient of friction, lubricant holding, and non-galling properties.
Hard nickel	Most ferrous and non-ferrous metals.	Softer than chromium but can be deposited faster and in greater thicknesses.
Rhodium	Most ferrous and nonferrous metals.	High hardness and wear-resistance combined with attractive appearance.
Porcelain enamels	Special enamelling sheet, cast iron, some aluminium castings.	Low friction coefficient, high abrasion resistance.
Flame sprayed coatings		
Metals	Almost all.	Wide variety of properties obtainable ranging from good bearing properties of soft nonferrous metals up to high hardness of high carbon steels.

Continued overleaf

TABLE 10.1—*Continued*

Type of coating or surface	Metals than can be coated	Major characteristics
Flame sprayed coatings (*contd.*)		
Ceramics and cermets	Almost all.	Wide range of hardness and wear-resistance, depending on coating material.
Hard anodic coatings	Most aluminium alloys and magnesium.	Oxide coating improves resistance to wear, abrasion and handling damage.
Hard facings	Most ferrous metals: not recommended with non-ferrous metals with melting point below 1100°C	Wide range of materials with hardness and wear-resistance properties to meet most conditions.
Heat treated surfaces		
Flame hardened	All hardenable steels: carbon steels and low alloy steels containing 0.35–0.7% carbon are preferred.	High hardness plus high strength and/or tough core. Can be used to harden selectively.
Induction hardened	Similar to those used for flame-hardened.	Generally same as flame-hardened.

10.2.2 Nitriding

Nitriding is one of the most effective processes for increasing hardness, wear-resistance, fatigue and corrosion resistance, and can be applied via the gas phase or in a salt bath at lower temperatures than are used for carburising. The hardness of typical nitrided surfaces is about 750/800 VPN and is maintained at temperatures to at least 620°C. The depth of the case ranges from 0.1 to 0.8 mm (0.005 in–0.030 in). Although distortion is low, the case formation is accompanied by some growth, but dimensional changes are predictable and can be allowed for by machining before nitriding.

A soft nitriding process named "Tufftriding" has had considerable success in the automotive industry. Surface layers of nitrides and carbides are formed on a metal surface giving good resistance to wear and galling.

TABLE 10.2 CASE DEPTH IN CARBURISED STEELS

Case Depth		Comments
<0.5 mm	<0.020 in	Wear-resistance at low loads
0.5–1.0 mm	0.020–0.040 in	Greater strength for moderate loads
1.0–1.5 mm	0.040–0.060 in	Used for severe abrasive, sliding or rolling wear combined with either heavy crushing loads or alternating bending stresses
>1.5 mm	>0.060 in	Used under very severe conditions have a tendency to distort and spall

10.2.3 Cyaniding and carbonitriding

Wear-resistant surfaces can be produced by incorporating both carbon and nitrogen into the case through liquid baths (cyaniding) and by the use of gas atmospheres (carbonitriding). Carbonitriding produces a case depth of 0.1 to 0.5 mm (0.003–0.020 in) with hardnesses approaching 800 VPN. Cyaniding produces thinner case depths of 6 μm to 0.25 mm (0.00025–0.010 in), with similar hardnesses.

10.2.4 Chromising

Wear-resistant chromium carbide cases can be produced by the diffusion of chromium into all high carbon steels, cast iron and many other metals. Hardness of the coatings ranges from 1600 to 1800 VPN (925 g load) and thickness from 10 to 50 μm (0.0005–0.002 in). Typical applications include guides, conveyor chain rollers and spiral hack-saw blades.

10.2.5 Siliconising

Substantial improvements to the wear resistance and hardness of steel and iron parts can be obtained by impregnating with silicon to form a case containing about 14% silicon. Carbon and sulphur contents of the base metal should generally be below 0.25 and 0.04%, respectively. Case depth can vary from 0.1 to 0.2 mm (0.0050–0.010 in), but the case tends to be rather brittle. The hardness varies from 150 to 160 VPN.

10.2.6 Sulphinuz

This is a sulphocyaniding process using an accelerated cyanide bath diffusing sulphur as well as carbon and nitrogen into the surface layer. It is used on steels and nickel-based alloys to confer good resistance to adhesive wear. It also has a high wear-resistance under impulsive loading and can be used for gears, clutch components, machine tools, etc.

10.2.7 Boronising

Boronising, a relatively recent innovation, is a diffusion treatment which, like carburising, can be carried out in a solid, liquid or gas medium. Although analogous to carburising, it produces a harder surface without recourse to quenching. The hardness of the surface layer is obtained directly through the formation of borides, surface hardnesses of 1200 to 1650 VPN being attainable. The depth of the hardened layer is dependent upon the diffusion time. The process has been applied to various steels and cast irons resulting in excellent sliding wear resistance.

10.3 ELECTROPLATES

The wear and abrasion resistance of many ferrous and non-ferrous metals can be improved considerably by electroplating them with hard or porous chromium, hard nickel, or rhodium.

10.3.1 Hard chromium

Hard chromium is popular for industrial wear applications because of its combination of hardness (1000–1025 BHN), low coefficient of friction (about 0.26 against cast iron), non-galling and non-wetting properties, and corrosion resistance. Depending upon the application, hard chromium electroplates are usually used in thicknesses of 1 to 300 μm (0.00005–0.012 in).

The hardness obtained depends upon the hardness of the base metal; full hardness is obtained on hard base metals such as hardened steel. Some uses and properties of hard chromium electroplates are given in Table 10.3.

TABLE 10.3 USES AND PROPERTIES OF HARD CHROMIUM ELECTROPLATES

Application	Hardness VPN 30 kg	Chromium thickness μm	Chromium thickness in
Drills	750–800	1–12	0.00005–0.0005
Reamers	750–800	2–12	0.0001–0.0005
Burnishing bars	700–750	12–75	0.0005–0.0030
Drawing plugs, mandrels	700–750	40–200	0.0015–0.0080
Drawing dies	750 inside 450 outside	12–200	0.0005–0.0080
Plastic moulds	600–700	5–50	0.0002–0.0020
Cages	485–650	2–40	0.0001–0.0015
Pump shafts	600–750	12–75	0.0005–0.0030
Rolls, drums	—	6–300	0.00025–0.0120
Hydraulic rams	—	12–100	0.0005–0.0040
Printing plates (engraved steel)	—	5–12	0.0002–0.0005

10.3.2 Hard nickel

Hard nickel is used principally in applications requiring a coating which is tougher and less liable to crack than chromium and where the extreme hardness of chromium is not essential.

Normally hardnesses of up to 650 VPN are possible but an increase to 800 VPN can be obtained by using nickel-phosphorus electrolyte followed by heat treatment. Deposits are usually limited to a maximum thickness of 6 mm.

10.3.3 Rhodium

Rhodium can be electroplated onto almost any base metal. Coating thicknesses of up to 5 μm (0.0002 in) are obtainable with hardnesses of 540 to 650 VPN. Applications usually make use of the good wear and corrosion resistance plus the attractive appearance of the coating.

10.4 PORCELAIN ENAMELS

Due to their high hardness and smooth surfaces, some grades of porcelain enamel can be used in applications requiring resistance to low-stress abrasion or a low coefficient of friction. Enamels have a Knoop hardness ranging from 150 to 560 and a glass-like surface.

Porcelain enamels can be applied to special grades of enamelling iron, cast iron and some aluminium castings. Although they are not recommended for high impact applications, they can be used in many wear applications where impact is not present—materials-handling equipment such as package chutes, bunker and silo discharge chutes and processing equipment for handling latex, tar products, etc.

Enamels also find an application where moderate abrasion resistance is required, e.g. coke, sand, gravel, ash and concrete handling chutes. The properties of heat and corrosion resistance exhibited by enamels are made use of in pump cylinders, burner grates, work surfaces and washing machine parts.

10.5 FLAME SPRAYED COATINGS

Sprayed coatings can be applied by several methods such as metallising (plating) and plasma arc and this is an economical way of applying wear and abrasion resistant coatings. They can be applied to almost all metals in thicknesses varying from a few microns to a few millimetres (0.0002–0.2 in). Generally, a sprayed coating is harder, more brittle and porous than the equivalent cast or wrought metal.

TABLE 10.4 ROCKWELL HARDNESS OF SPRAYED METAL COATINGS

Aluminium alloy	H72	Stainless steel (13.5% Cr)	C29
Aluminium bronze	B78	Steel	
Copper	B32	0.10 C	B89
Iron	B80	0.25 C	B90
Molybdenum	C38	0.80 C	C36
Monel	B39	Tin	H10
Nickel	B49	Tobin-type bronze	B50
Stainless steel (18–8 ELC)	B78	Zinc	H46

Tables 10.4 and 10.5 show typical relative hardnesses of metals deposited by metallising and plasma arc spraying. Tables 10.6 and 10.7 give further and more detailed properties and applications.

Detailed properties of some of the coatings listed in Table 10.6 are shown in Table 10.7.

10.6 HARD ANODIC COATINGS

Hard anodic coatings are most commonly used on aluminium but can be applied to other non-ferrous metals such as zinc and titanium. Both conventional and hard anodic coatings on aluminium are composed of oxide and therefore have similar hardnesses. The hard anodic coating, however, has better resistance to abrasive wear and corrosion because of its greater thickness, density and weight. Hard

TABLE 10.5 VICKERS HARDNESS OF TYPICAL PLASMA SPRAYED COATINGS

Metals		Carbides and oxides	
Molybdenum wire		Hafnium carbide	1000
as sprayed	168–183	Tantalum carbide	1000
heat treated[1]	155–159	Osmium oxide	168–205
Molybdenum powder		Chromic oxide	1000
as sprayed	321–368	Chromium-aluminium oxide	287–455
heat treated[2]	168–193	Hafnium oxide	251–313
Tantalum powder		Zirconium oxide	638–805
as sprayed	443		
heat treated[3]	490		
Tungsten wire			
as sprayed	246–263		
heat treated[4]	238–255		
Tungsten powder			
as sprayed	330–390		
heat treated[3]	162–200		

[1] In hydrogen at 3600F. [2] In hydrogen at 2150F. [3] In vacuum at 4000F. [4] In hydrogen at 3600F.

TABLE 10.6 PROPERTIES AND APPLICATIONS OF SOME SPRAYED COATINGS

Stainless steel (13.5% Cr)	—hard coating, high strength and elongation, low shrinkage and cracking tendencies. Used for armature shafts, journal surfaces, cylinder liners, pistons, valve stems, plungers and rams.
High carbon steel	—hard wear-resistant coating, similar applications as for stainless except where corrosion resistance is also required.
Molybdenum coatings	—good wear and abrasion resistance, good adhesion. Used to build up worn parts and also for journal surfaces.
Copper alloys	—general purpose wear applications.
Self-fluxing alloys	—Cr/Ni alloys plus fluxing agent. High hardness with high abrasion wear-resistance.
Tungsten carbide	—wide composition range, flame spraying or flame plating. Spray can achieve hardnesses of ~750 VPN for the matrix with 1200+ VPN for the carbides. Used for severe abrasive conditions. Plating used for wear-resistance plus shock, heat or corrosion resistance.
Aluminium oxide	—popular for wear and heat applications. Superior to tungsten carbide under corrosive or high temperature oxidising conditions, but inferior to WC in dry running conditions.

TABLE 10.7 PROPERTIES OF FLAME-PLATED COATINGS

Coating →	Tungsten carbides WC+6.8% Co	WC+13.15% Co	WC+15.17% Co	25% WC + 7% Ni + mixed W–Cr carbides	85% Chromium carbide (Cr_3C_2) +15% Ni–Cr	99+ % Aluminium oxide (gamma)
Mechanical properties						
Vickers hardness	1200–1450	1100–1250	1050–1200	1000–1200	1000–1200	1000–1200
Mod. of rupture, psi	67 000–73 000	80 000–90 000	90 000–100 000	40 000	75 000	22 000
Mod. of elast., 106 psi	40	36–38	34–36	17	22	15
Porosity, %	0.5	0.5–1	0.7–1.2	0.5	0.5	1
Thermal properties						
Max. cont. temp., F	1000	1000	1000	1400	1800	1200–1800*
Coef. of ther. exp. per °F	4×10^{-6} (70–1000F)	4.5×10^{-6} (70–1000F)	4.7×10^{-6}	4.6×10^{-6} (70–1400F)	6.4×10^{-6} (70–1800F)	3.9×10^{-6} (70–1800F)
Spec. ht, Btu/lb °F	0.048	0.054	0.056	0.07	0.127	0.196
Ther. cond., Btu/hr/ft^2 °F/ft	5.3(70–500F)	5.3(70–500F)	5.3(70–500F)	3.8(500F)	4.3(500F)	0.85(500F)
Main features	Extreme wear-resistance.	Excellent wear-resistance plus increased resistance to mechanical and thermal shock.	Excellent wear-resistance plus greatest resistance to mechanical and thermal shock.	Excellent wear-resistance at higher temperatures. Improved corrosion resistance.	Good wear-resistance at high temperatures or in corrosive media. Resists flame impingement.	Excellent resistance to wear, chemical attack and high temperature oxidation.

* Maximum service temperature of aluminium oxide is highly dependent on base metal used and on type of service.

anodised coatings have a hardness of about 400/500 VPN and thicknesses of up to 0.1 mm (0.004 in), which imparts good abrasion resistance. The major user of hard anodic coatings is the aircraft industry, for applications requiring light weight and high wear-resistance, such as undercarriage parts, control cylinders, cams and tracks, servo valves and pistons.

10.7 HARD FACINGS

Two methods of hard facing are commonly used. One uses an oxyacetylene torch and a welding rod composed of the metal which is to be applied to the surface. In other words, the facing metal is welded to the base metal. The second uses a metallic arc welding machine and the facing metal is the welding electrode. Typical wear-resistant hard facings are shown in Table 10.8, with a general guide to the selection of hard facing alloys for particular wear applications in Table 10.9.

Hardfacing can be applied to most ferrous metals but is not recommended with non-ferrous metals having melting points below 1100°C. The thickness of the facing varies with the application but is usually in the range of 1.5 to 6.0 mm.

Carbon steels are relatively easy to hard face, but with increasing carbon content welding becomes more difficult and heat treatment

TABLE 10.8 A GRADED SERIES OF WEAR-RESISTANT HARD FACINGS

Increasing abrasion resistance ↑ / Increasing toughness ↓	Hard facing	Properties
	1. Tungsten carbide	Maximum abrasion resistance: worn surfaces become rough.
	2. High-chromium irons	Excellent erosion resistance: oxidation resistance.
	3. Martensitic irons	Excellent abrasion resistance: high compressive strength.
	4. Cobalt-base alloys	Oxidation resistance: corrosion resistance: hot strength and creep resistance.
	5. Austenitic irons	Good erosion resistance.
	6. Nickel-base alloys	Corrosion resistance: may have oxidation and creep resistance.
	7. Martensitic steels	Good combination of abrasion and impact resistance: good compressive strength.
	8. Pearlitic steels	Inexpensive: fair abrasion and impact resistance.
	9. Austenitic steels: manganese steel	Work-hardening: maximum toughness with fair abrasion resistance: good metal-to-metal wear-resistance under impact.

TABLE 10.9 A GENERAL GUIDE TO THE SELECTION OF HARD-FACING ALLOYS

Service conditions	Preferred materials
Adhesive wear	
a. Oxidative or mild wear	Iron-base alloy 6 to 16% Cr; Co or Ni-base alloys 15 to 30% Cr
b. Metallic or severe wear	Cobalt or nickel-base alloys
Abrasive wear	
a. Low stress scratching abrasion	Carbide grains in alloy iron, cobalt or nickel High chromium irons 3.5 to 4.5% carbon
b. High stress grinding abrasion	High chromium martensitic irons 2.0 to 3.5% C Martensitic alloy steels 0.4 to 1.5% C
c. Gouging abrasion (plus impact)	Austenitic manganese steels with High chromium irons, 2.0 to 3.5% C
Erosion	
a. Low angle impingement	Carbide grains in alloy iron, cobalt or nickel High chromium irons 3.5 to 4.5% C Hyper-eutectic cobalt- or nickel-base alloys
b. High angle impingement	High chromium irons 2.0 to 3.5% C Hypo-eutectic cobalt-base alloys
c. Cavitation-erosion	Hypo-eutectic cobalt-base alloys
Fretting corrosion	Cobalt- or nickel-base alloys
Corrosion-erosion	Cobalt or nickel-base alloys High chromium steels 12–18% Cr, 0.2–1.2% C
Oxidation-hot corrosion	Laves phase Ni or Co-base alloy Carbide containing Ni or Co-base alloys
High temperature wear	Cobalt-base alloys Laves phase cobalt or nickel-base alloys

must be carried out prior to and after hard facing of high carbon and low alloy steels. High carbon, stainless steels, cast iron, ductile iron and high speed steels can be hard-faced provided due precautions are taken, as can Monel, but copper and its alloys are more difficult due to their low melting points and high thermal conductivity. The wear-resistant properties of various hard facing materials to gouging abrasion are shown in Fig 10.1.

10.8 NEW COATING TECHNOLOGIES

In the past decade new coating technologies have been developed which offer considerable flexibility and process economies in the modification of surface properties to resist wear. The new

V.P.N.

V.P.N.	Material
700	Composite tungsten carbide–arc weld
800	Martensitic iron–gas weld
640	Martensitic iron–arc weld
550	Cr–Co–W #1–gas weld
675	Martensitic steel–electric weld
600	Austenitic iron–gas weld
200	Austenitic Mn steel–electric weld
340	Pearlitic steel–gas weld
107	SAE 1020 steel (standard)
475	Cr–Co–W #6–gas weld

Increasing abrasion resistance ⟶

FIG. 10.1 RESISTANCE OF SOME FACING MATERIALS TO GOUGING WEAR

technologies, which include pyrolytic decomposition, chemical vapour deposition (CVD), physical vapour deposition, reactive evaporation and sputtering, involve the deposition of a wide range of soft and hard compounds and metals.

The tribological applications of the new coating technology extend over a wide range; however, the following are some of the major uses.

10.8.1 Bearing applications

Adherent films of molybdenum disulphide on bearing surfaces used in the space industry have been developed by sputtering techniques (DC and RF). Since the coating thickness is usually small, typically 0.2–2.0 μm, components can be coated as a final processing stage. The corrosion resistance or endurance of the coating can be improved by the use of thin underlay.

Soft coats are possible with metals such as gold, silver and lead. However, ion plating, due to greater adhesion, has been found to be superior to vacuum evaporation for soft coats when used under high vacuum conditions. Rolling element bearings for high-vacuum service can also be produced with hard coats of thin layers of CVD TiC, but suffer the disadvantage of substrate structure changes which can occur when using high temperature CVD techniques. The sputter-coating of hard materials (TiC, Mo_2C, Mo_2B_5 and $MoSi_2$) can avoid any substrate structural changes associated with CVD, thus improving considerably the wear life of the component.

10.8.2 Wear-resistant coated cutting-tools

A wide range of CVD techniques is used to produce thin hard coats of carbides, nitrides, carbo-nitrides and oxides on cemented carbide substrates. Coating thicknesses of typically 4 to 7 μm are used. High-speed steel tools can be coated by RF sputtering to avoid the risk of over-tempering when using conventional CVD processes.

Hard coats of metal carbides can be produced also by activated reactive evaporation or by ion plating followed by induction-hardening in selected atmospheres.

10.8.3 Metal working tools

Boriding of deep drawing tools, mandrels and stamping dies has been found to give increased wear life. Complex shapes such as screws are suitable for hard coating by CVD techniques. Fabrication to shape on selected preforms can also be obtained using CVD, e.g. patterns for mould production in metal casting.

10.8.4 Erosion-resistant coatings

Hard coats produced by CVD have been found to increase the erosion resistance of compressor blades in turbine engines. The erosion life can be some 70 to 80 times greater than that of plasma-sprayed blades. CVD coatings of tungsten or tungsten/tantalum have been applied to gun barrels in order to overcome the frequent failure of the hard chrome plate normally used. Such coatings produce a markedly lower erosion rate.

10.8.5 Coatings for non-ferrous metals

Metallic soft coats are normally used, e.g. ion plating of soft metals, gold, silver, cadmium and lead; also coating of chromium and zirconium have been found to improve wear-resistance. Hard coats, although less common, have been used on aluminium and copper alloys using the technique of sputter deposition.

10.9 HEAT-TREATED SURFACES

Wear-resistant surfaces are obtained on many steels and iron castings by either flame-hardening or induction-hardening. Both methods use the same principle of rapidly heating the surface to a temperature high enough to permit hardening by quenching.

10.9.1 Flame-hardening

The case depths and hardnesses obtained vary according to the heating and cooling rates employed. Depth of hardening ranges from

3 to 6 mm and as no change in chemical composition occurs there is a gradual reduction in hardness away from the surface until the original hardness is reached.

The process is applicable to all hardenable steels, low alloy steels, hardenable stainless steels and tool steels. Grey cast irons can also be flame hardened but in order to stabilise the iron and prevent graphitisation, alloying with small amounts of chromium or molybdenum is required. Pearlitic malleable and nodular cast irons are also hardenable but care must be taken to prevent cracking by stress relieving before and after hardening.

10.9.2 Induction-hardening

A greater speed of heating is obtained with induction-hardening than with flame-hardening: it is therefore possible to produce thinner cases. The steels used for induction-hardening are similar to those for flame-hardening. Nodular irons can be induction-hardened (to 600/700 VPN) if they are stress relieved and have the correct structure. Pearlitic malleable iron and grey iron can also be treated.

Chapter 11

Effects of Lubricants on Wear

11.1 INTRODUCTION

A lubricant has been defined as "Any substance interposed between two surfaces in relative motion for the purpose of reducing the friction and/or wear between them". In fact, many solids or liquids can be effective lubricants which are not interposed for that purpose, such as wet clay on a ploughshare, the liquid component of a process slurry or even fine dust in a milling process.

Obviously the effect of a lubricant is usually to reduce wear, but sometimes wear will be increased by a lubricant and there are situations in which wear is unchanged, or where the type of wear is changed but not the amount of wear. The object of this Chapter is therefore to give a brief picture of some of the ways, desirable or undesirable, in which a lubricant can affect wear.

A lubricant may reduce friction and not wear, or wear and not friction, but in practice most lubricants are intended to do both. There are two ways in which a lubricant can reduce wear, either directly or by virtue of a reduction in friction. Conversely there are four ways in which a lubricant can increase wear or change it to a more undesirable form. These are by encouraging a change from mild to severe wear in a two-body wear situation by interfering with oxidation, by being itself degraded to a harmful form, by chemical attack (corrosive wear), or most important by helping to trap abrasive particles and form an abrasive paste or slurry.

11.2 INDIRECT EFFECT BY REDUCING FRICTION

The general effect of a lubricant in reducing friction is best explained by reference to Fig. 11.1 which shows the relation between friction and the Sommerfeld Number for a journal bearing.

The Sommerfeld Number is ZN/p where Z is the lubricant viscosity, N the rotational speed of the bearing and P the nominal pressure on the bearing surface.

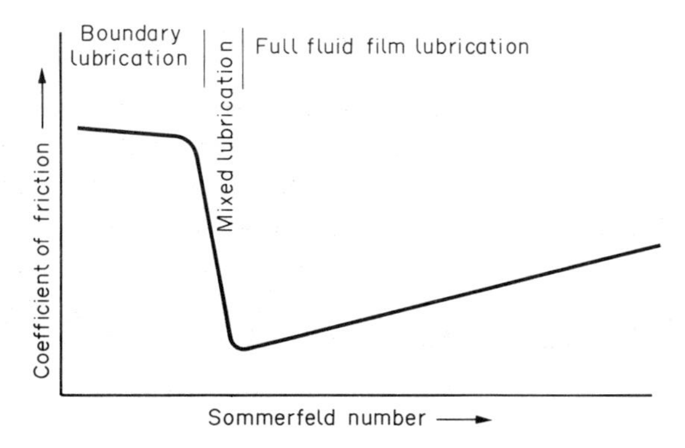

FIG. 11.1 RELATION BETWEEN FRICTION AND SOMMERFELD NUMBER

Friction represents energy lost or dissipated in the contact region. The higher the friction the greater is the energy dissipated, and therefore the higher the temperatures at the contact.

In the region of full fluid film lubrication there is little or no solid–solid contact, the friction is low, and there is very little or no wear. In the mixed and boundary lubrication regions there is increasing friction as the Sommerfeld Number decreases, and the energy dissipated increases. There is therefore increasing heat generated as lubrication becomes less effective until the point is reached where no effective lubrication exists and the contact is effectively unlubricated. The above discussion is based on a plain journal bearing, but the basic argument applies to all lubricated contacts.

It follows that the effect of a lubricant in reducing friction also reduces the heat dissipated and therefore the temperatures in the contact region, and the degree of temperature reduction improves as lubrication becomes more effective.

The advantage of temperature reduction in relation to wear is threefold. Where there is a tendency for adhesive wear, it is greater at higher temperatures. A reduced temperature therefore reduces the incidence of adhesive wear. Secondly, where there is abrasion of a softer material by a harder one, the ratio of hardnesses will often become worse at higher temperatures, and the wear of the softer material will increase. Reducing the temperature will therefore tend to reduce abrasive wear. A third important factor is that fatigue or loss of adhesion with layered materials will often be worse at higher temperatures.

11.3 DIRECT REDUCTION OF WEAR BY A LUBRICANT

Even in the boundary lubrication region a lubricant can interpose a film between two solid surfaces and thus reduce asperity contact and wear. The effect is particularly strong in reducing adhesive wear, and there may be a 99% reduction in wear rate, but the degree of reduction depends on the effectiveness of film formation.

Reactive materials may be added to the lubricant specifically to reduce wear. One type, called an anti-wear additive, adsorbs on a surface to form a film which is stronger than that of the liquid lubricant. It therefore gives better resistance to penetration and reduces wear under fairly mild conditions. A second type, called extreme pressure additives, apparently reacts with metal surface which has been exposed by wear and prevents it from welding to a similar exposed surface on the counterface. In this way it prevents severe tearing of the contact surfaces.

Liquid or grease lubricants are not usually very effective under fretting conditions, but dry lubricants, especially those based on molybdenum disulphide, are often useful.

11.4 REDUCTION OF OXIDATION

It is probable that the transition from mild to severe wear in two-body rubbing takes place when the protective oxide film is not adequately repaired after wear has occurred. It is possible for some lubricants to interfere with the regeneration of the oxide film and thus to promote transition from mild to severe wear, but in practice it is doubtful whether this is a serious problem with conventional lubricants.

11.5 DEGRADATION OF LUBRICANT

Lubricants which are subjected to excessively high temperatures will be degraded, and there are several ways in which the degraded lubricant can contribute to increased wear problems.

The mildest form of harmful degradation product is a resinous or lacquer-like material which can cause increased friction or stick-slip friction. These materials can lead to higher temperatures and unsteady loading, and thus to increased wear or bearing damage.

More severe degradation of lubricants, especially of inorganic materials such as silicones or some grease thickeners, can produce abrasive solid degradation products which directly produce abrasion of sliding surfaces. Although there have been reports of this occurring, it is far from certain that it represents an important practical problem.

Situations sometimes occur in sliding where extremely high temperatures arise, as high as 1000°C or more, and some systems have survived such high intermittent temperatures without seizure or serious damage in the absence of lubricants. Where small amounts of lubricant are present in such situations, however, the lubricant can act as a carbon source to produce carburisation of a steel surface. The resulting very hard rough region can then produce severe bearing damage.

11.6 CHEMICAL ATTACK ON A SURFACE

Lubricants often contain materials which react with a metal surface, and the surface films so formed are commonly more easily abraded away than the metal itself or its oxide. Thus, under mild abrasion conditions the wear rate can be increased by this mild corrosive attack.

Extreme pressure additives are particularly likely to cause increased wear under mild abrasive conditions because they are powerful chemical reagents, but water in a lubricant can have a similar effect.

11.7 TRAPPING ABRASIVE PARTICLES

The worst deleterious effect of lubricants is in trapping loose abrasive materials and preventing them from being swept out of the contact region. The abrasives are then subjected to repeated loading and may fracture to produce sharper asperities. The overall effect is the formation of a sort of grinding paste which can cause severe grinding wear. On a small scale this behaviour may be responsible for the relative ineffectiveness of liquid or grease lubricants in reducing fretting.

Where abrasive dusts, particles or lumps are present there are three possible approaches to the problem of lubrication.

(a) Leave the system unlubricated. With dry surfaces abrasive materials will tend to be swept away from the contact region.

(b) If lubrication is essential, use a powerful oil flow which will also sweep away abrasive materials, and use a separation or filtration system to remove abrasive materials from the oil.

(c) If neither of the above approaches can be used, consider the use of a dry lubricant. This can reduce friction in the contact region, but will not usually reduce abrasive wear as compared with the unlubricated situation.

11.8 SUMMARY

Although this chapter has described disadvantages of lubrication more than its advantages, it will be obvious that, in general, lubricants tend to make wear situations better rather than worse. The one importance exception is in the presence of loose abrasive materials, where the use of liquid or grease lubricants should be avoided if possible.

Chapter 12

Wear Testing

12.1 WEAR TESTING

In 1964 the O.E.C.D. Group on Wear of Engineering Materials initiated a detailed examination of the correlation between different wear test methods. Their main conclusions were: that in laboratory testing the absolute values of friction and wear must be treated with extreme care; that if only one wear mechanism occurs the reproducibility of the results is very reasonable; and that seemingly minor factors, especially thermal conditions at the interface, may have a considerable influence.

They recommended that "When materials are to be classified with respect to their wear-resistance, the utmost care should be taken that the test rig is constructed or modified in such a way that the practical conditions, especially those that relate to temperature and environment, are approximated as closely as possible. Even then some materials on which practical information is available should be included in the programme as reference materials".

The ideal performance test will therefore be a service trial, but there are several factors which may make service testing unsatisfactory.

(a) *Duration of trial:* the service life may be so long that results from a service trial will not be available quickly enough to be useful.
(b) *Lack of control:* it is usually not possible in a service trial to control all the operating or environmental conditions, such as speed, load, temperature and humidity, or to vary them over the whole of the specified range.
(c) *Inadequate monitoring:* service equipment will often not permit adequate monitoring of all the important factors, such as temperatures, loadings and torques. Similarly the components may not be available for examination after a service trial.

The cost of a service trial may also be excessive, particularly where the fabrication of single components from experimental materials is necessary. However, this is not always true, and a fully controlled laboratory test may be more expensive than a service trial.

A second level of simulation may be obtained by testing the actual service equipment in a laboratory, using either accurately simulated or even exaggerated operating conditions. This approach has the advantage of a very high level of control and monitoring, but the cost of such testing is likely to be high.

In practice, because of the limitations of service testing and the cost of laboratory testing of service equipment, most wear testing is carried out on laboratory rigs which attempt to simulate some or all of the service conditions.

12.1.1 Laboratory testing

In general, the reliability and relevance of the results will be greatest when the highest simulation of the service conditions is used. If for any reason a test condition differs from the service condition, it is important to ensure that there is no change in the nature of the wear process between the two sets of conditions which would make the test results valueless.

For example, it is common for the test specimens used in laboratory rigs to be smaller than the actual service components, and the contact area in the wear test may also be scaled down. If there is much energy dissipated in the wear contact, then the temperatures throughout the test section will not be uniform, and the heat transfer characteristics of the smaller test specimen will differ from those of the service components. The actual contact temperature will therefore tend to be different.

In such a case it will be important to establish that there is no marked change in the wear behaviour at temperatures similar to those occurring in the test or the service application. This can be done by testing at a series of temperatures broad enough to include the full relevant range.

One difficulty in laboratory wear testing is to accelerate a test without making the result unrealistic, and care must always be taken to ensure that increased loads, speeds or temperatures produce only the desired acceleration of the test without completely altering the behaviour of the material or the nature of the wear process.

A particularly important aspect of this problem is the use of non-conformal contact geometry to accelerate wear testing. Most practical wear situations arising between two rubbing surfaces involve more or less conformal contact in which the wear rate will be relatively low. It is common in laboratory testing to use a non-conformal line or point contact in order to produce more rapid wear. Unfortunately the use of the results of such a test to predict service wear will be less reliable, and it is even possible for such tests to alter the relative rating of two materials.

Having selected the appropriate test conditions, arrangements should be made for the amount of wear to be measured periodically during the test. This can be done conveniently but with limited accuracy by measuring the linear movement of a specimen holder loaded against a fixed counterface. A more accurate technique for small wear scars is to measure the dimensions of the scar with a microscope and/or a profilometer such as the Talysurf.

Probably the most accurate technique is to weigh the specimen at regular intervals, but this involves designing the specimen so that its total weight is low enough for the weight loss to be measured accurately (e.g. a weight loss greater than 0.0005 g against a total weight of 100 g). It also involves a mechanism and procedure for removing and weighing the specimen and replacing it without disturbing the wear process.

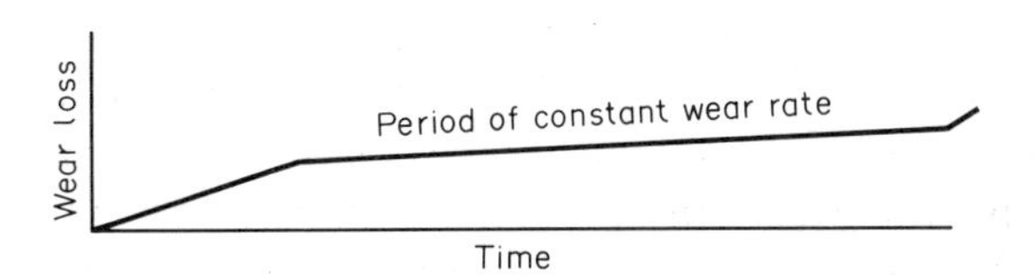

FIG. 12.1 TYPICAL PLOT OF WEAR AGAINST TIME

A typical plot of wear against time is shown in Fig. 12.1. The initial wear rate is not constant but varies as such factors as surface finish and hardness are changing. There then follows a relatively long period during which the wear rate is constant, and it is on this steady wear period that material selection and wear life predictions should normally be based.

To summarise, the object in laboratory wear testing should be to simulate the service conditions as closely as possible. Where circumstances make it necessary for the test conditions to differ from the service conditions, a range of test conditions should be used to establish that the results can be reliably extrapolated. Finally, the selection of materials and prediction of wear life should normally be based on the period of constant wear rate in the test history.

12.1.2 Test rigs

There are two different approaches to the design or selection of laboratory test rigs, namely to design a rig to study a specific wear situation, or to use one or more standard rigs.

The use of a specially-designed rig will usually permit a closer simulation of the required service conditions. The cost will often be intermediate between that of testing the service equipment and that of a standard wear test rig, but this is not always the case, and the cost should always be assessed for the different test approaches for any particular wear study.

The main disadvantage of a specially-designed wear test rig is that it will be restricted in the range of situations which can be simulated, and therefore in the range of applicability of the results.

The use of a standard wear test rig will usually only permit a much lower degree of simulation of the service conditions. The cost will normally be relatively low, both in the initial capital cost of the rig because there will be no design or development costs, and in the operating costs because the test specimens will usually be simple and mass-produced.

Where a user is only concerned with one wear situation, or a restricted range of wear situations, it may be more effective and economical for him to use a purpose-built test rig. If he has to examine a wide range of wear situations it may be better to use a range of standard commercial test rigs, each capable of simulating some part of the relevant range of service conditions.

Table 12.1 lists the characteristics of a number of commercially-available wear test rigs, but many of these are intended primarily for assessing lubricated situations. Most of them can be used for unlubricated tests, but only a few such as the LFW-3, Cygnus, LFW-1 and Reichert can be effectively used for assessing two-body wear, and none is suitable for studying the effect of abrasive particles, except possibly the LFW-3 and Coturnix. Figure 12.2 shows two techniques which have been used to study grinding or gouging abrasion.

12.2 APPLICATION OF WEAR TEST RESULTS

12.2.1 Selection of materials

If wear tests have been carried out with a high degree of simulation of the service situation, then the results can be used with considerable confidence in selecting the best wear-resistant materials.

Where the test simulation was less good, there will be a greater risk that the best material has not been identified.

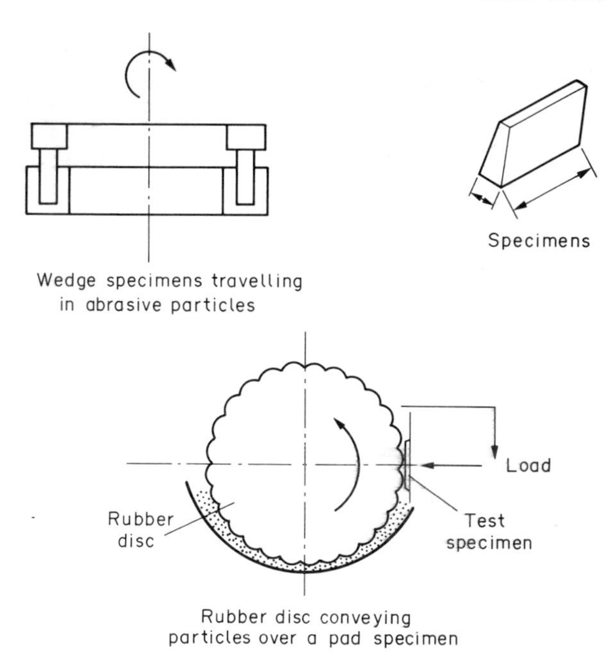

FIG. 12.2 RIGS USING ABRASIVE PARTICLES

In any case a service trial should preferably be carried out, with periodic examination to the components to ensure that the wear behaviour is acceptable. On the basis of the wear test results two or more materials might be selected for service trials, in order to ensure that the same relative rating applies in service.

It is often critically important to ensure that the material used in service is identical with that used in the wear tests, in composition, treatment, microstructure and surface quality. This is particularly true of cast materials, where the casting of a larger component may introduce differences of composition and structure sufficient to completely destroy the relevance of the tests.

12.2.2 Prediction of wear life

Prediction of wear life from laboratory tests is generally poor. The method most commonly used is to calculate a wear coefficient K, where

$$K = \frac{\text{Volume of material removed by wear}}{\text{Load} \times \text{sliding distance}}$$

and is usually reported in $mm^3/N\ m$. The use of this coefficient is based on the assumption that the volume wear varies directly with the

TABLE 12.1 CHARACTERISTICS OF COMMERCIAL LUBRICANT AND WEAR TEST RIGS

			Measurement					Conditions			
Contact	Test machine	Geometry	Static friction	Kinetic friction	Wear	Wear life	Load capacity	Maximum load (N)	Speed (max) (rpm)	Speed (max) (m/s)	Temp. (max) °C
Conformal, area sliding	Alpha LFW-3	Annular ring, flat block	•	•	•	•	—	2.2×10^4	325	0.26	650°
	Alpha LFW-3	Pin in bush	•	•	—	—	—	—	—	—	Ambient
	Almen-Wieland	Pin in half shells	—	•	•	•	—	1.9×10^4	600	0.2	Ambient
	Cygnus	Pin on disc or flat washers	•	•	•	•	—	300	3000	25	150°
	Coturnix	Journal bearing	•	•	•	•	—	10^4	3000	16	200°
Non-conformal line, sliding	Timken	Block on ring	•	•	—	•	•	4.4×10^3	800	2.0	Ambient
	Falex	Pin in V-blocks	—	•	•	•	•	2.0×10^4	750	0.25	Ambient
	Alpha LFW-1	Block on ring	•	•	•	•	•	2.8×10^3	200	0.4	Ambient
Non-conformal point, sliding	Shell 4-ball	Ball on ball	—	•	•	•	•	1.8×10^4	1500*	0.5	200°
	Reichert	Crossed cylinders	—	•	•	—	—	580	900	—	Ambient
Non-conformal point rolling	Shell 4-ball	Ball on ball	—	—	•	•	•	1.8×10^4	1500*	0.5	200°

* 1800 rpm for American machines.

contact load and the sliding distance, or alternatively that the wear depth varies directly with the contact pressure and the sliding distance. An alternative form of this relationship is

$$L = \left(\frac{S_r}{S}\right)\left(\frac{P_r}{P}\right)L_r$$

where L and L_r are the predicted and test lives, S and S_r the service and test speeds, and P and P_r the service and test loads, but these relationships are only valid within limited ranges of speed and load.

The product PV of contact pressure and sliding speed is often used to define the wear behaviour of a material. There is an upper limit to PV for recommended use, above which the wear rate increases rapidly. This will sometimes give a better prediction than the use of a simple wear coefficient.

In practice the prediction of wear life by any of these techniques on the basis of laboratory tests is poor. Provided the degree of test simulation has been good the most accurate prediction may be based on a comparison of wear rates for two different materials. Thus if in a laboratory test a new material B has a wear rate half that of an existing material A, it will be reasonable to expect that B will give a service wear life of the order of twice that of A.

It must be borne in mind that all of these relationships will only apply to homogeneous materials. Where a component has a hardened surface the wear rate may be quite different once the hardened layer has been penetrated.

12.3 SUMMARY

In conclusion, where wear testing has been carried out with very good simulation of the service conditions, the selection of materials for service use can be made with a high level of confidence, and a useful prediction of service life can be made. This is particularly true when a comparison can be made with a material which has already been used in service.

Where the simulation of service conditions was less good, there will be less confidence in the selection of materials, and especially in the prediction of wear life, and some controlled assessment of service performance will be needed.

Suppliers of Materials

FERROUS METALS

Key to types of steel marketed

a = Carbon steels
b = Alloy steels
c = High-strength steels
d = Stainless steels
e = Heat-resisting steels

---

a,b,c British Rolling Mills Ltd., Brymill Steel Works, P.O. Box 10, Tipton, Staffs. Tel: 021 557 3939

a,b,c,d Carr & Co. Ltd., Richard W., Pluto Works, Wadsley Bridge, Sheffield S6 1LL. Tel: 0742 349451

a,b,c Caxton Steel Co. Ltd., Cumberland Avenue, London N.W. 10. Tel: 01 965 5700

a,b,d Dobell Coated Steel, Scandia Steel, Hayes Lane, Lye, Nr. Stourbridge, Worcs. Tel: 038482 6061

a,b,c,d Doncasters Sheffield Ltd., Penistone Road, Sheffield S6 2FR. Tel: 0742 349444

a,b,c,d Dunlop & Ranken Ltd., Whitehall Road, Leeds LS1 1HB. Tel: 0532 636363

a,b,c,d Firth Brown Ltd., Atlas Works, Sheffield S4 7US. Tel: 0742 20081

a,b,c Firth Co. Ltd., The, Warrington, Lancs. Tel: 0925 38881

b,c,d Firth-Vickers Stainless Ltd., Staybrite Works, Weedon Street, Sheffield S9 2FU. Tel: 0742 449955

a,b,c Furnival Steel Co. Ltd., Attercliffe Road, Sheffield 4. Tel: 0742 20403

b,c,d,e Hall & Pickles Ltd., P.O. Box 161, Ecclesfield, Sheffield S30 3ZF. Tel: 0742 467131

a,b,d Hobson Houghton & Co. Ltd., Savile Street, Sheffield 4. Tel: 0742 41245

b,c,d Ireland Alloys Ltd., P.O. Box 18, Hamilton, Scotland. Tel: Blantyre 0698 822461

a,b,c,d Lee & Sons Ltd., Arthur, P.O. Box 54, Trubrite Steel Works, Meadow Hall , Sheffield S9 1HU. Tel: 0742 387272

a,b,c,d Macready's Metal Co. Ltd., Usaspead Corner, Pentonville Road, London N1 9NE. Tel: 01 837 7060

a,b Moss & Gamble Bros. Ltd., Mowbray Street, Sheffield 3. Tel; 0742 20178

a,b,c Sanderson Kayser Ltd., P.O. Box 6, Newshall Road, Sheffield S9 2SD. Tel: 0742 449994

b,c,d Sandvik UK Ltd., Manor Way, Halesowen, Worcs. Tel: 021 550 4700

b,c,d Simpson Ltd., Alfred, Bridge Street, Swinton, Manchester M27 1EL. Tel: 061 794 4777

CAST IRON FOUNDERS

Key to types of iron cast:

a = Grey iron
b = Spheroidal-graphite (SG) iron
c = Meehanite
d = Nickel iron

---

a Adamson-Alliance (Horsehay) Ltd., Horsehay, Telford, Salop. Tel: 0952 505881

a Babcock & Wilcox (Operations) Ltd., 165 Great Dover Street, London SE1 4YB. Tel: 01 407 8383

a Barton Conduits Ltd., Old Birchills, Walsall WS2 8QE. Tel: 0922 26581

b Blakeborough & Sons Ltd., J., P.O. Box 11, Brighouse HD6 1NH. Tel: 0484 715511

a,b British Steel Corp. (General Steels and Special Steels Divs)., 33 Grosvenor Place, London W.C.1. Tel: 01 235 1212

a Brockhouse Castings Ltd., Hall Street, Wednesfield, Wolverhampton. Tel: 0902 731221

a,b,c Dorman Diesels Ltd. (English Electric Diesels Group), Tixall Road, Stafford, Staffs. Tel: 0785 3141

a,b Duport Foundries Ltd., Dudley Port, Tipton, Staffs. Tel: 021 557 3963

a,c Eurocast Bar Ltd., John, Britannia Foundry, Meadow Lane, Loughborough, Leics. Tel: 0509 212632

a Firth Brown Ltd., P.O. Box 114, Atlas Works, Sheffield S4 7US. Tel: 0742 20081

d Follsain-Wycliffe Foundries Ltd., Lutterworth, Rugby. Tel: 045 55 3551

a,b G.K.N. Engineering Ltd., P.O. Box 19, Redditch, Worcs. Tel: 0527 25222

b Harland & Wolff Ltd., Queen's Island, Belfast BT3 9DU. Tel: 0232 58456

b Head Wrightson Iron Foundries Ltd., P.O. Box 10, Teesdale Works, Thornaby-on-Tees, Teesside. Tel: 0642 62241

b,c International Meehanite Metal Co. Ltd., Albert Road, North Reigate, Surrey. Tel: 073 72 44786

b Lloyd & Co. Ltd., F.H., P.O. Box 5, James Bridge Steel Works, Nr. Wednesbury WS10 9SD. Tel: 021 526 3121

a,b,d Moyle Partnership, The Victor, 73 Walton Road, Woking, Surrey. Tel: 04862 4387

a,d Qualcast Ltd., Victoria Road, Derby. Tel: 0332 23260.

a Saunders Valve Co. Ltd., Grange Road, Cwmbran, Newport, Mon. Tel: 06333 2044

a Triplex Foundry Ltd., Upper Church Lane, Tipton, Staffs. Tel: 021 557 1293

a Wedge & Wright (Engn.) Ltd., Knottingley, Yorks. Tel: 0977 82743

NON-FERROUS METALS

Key to types marketed

a = Aluminium & its alloys
b = Copper & copper base alloys
c = Nickel & its alloys
d = Tin & its alloys
e = Titanium & its alloys
f = Refractory metals
g = Niobium & its alloys
h = Tungsten & its alloys
j = Cemented carbides

---

a,b,e Analco Ltd., High Road, Cowley Peachley, Uxbridge, Middlesex. Tel: 08954 46511

a Baco Aluminium (Ravensbourne) Ltd., 512 Purley Way, Croydon, Surrey. Tel: 01 686 8451

b Thomas Bolton & Sons Ltd., P.O. Box 1, Froghall, Stoke-on-Trent. Tel: 05384 2241

a,b,d Bush Beach Engineering Ltd., Stanley Green Trading Estate, Cheadle Hulme, Cheadle, Cheshire SK8 6RN. Tel: 061 485 8151

b,d Delta Metal Co. Ltd., 1 Kingsway, London WC2B 6XF. Tel: 01 836 3535

e Deloro Stellite (UK) Ltd., Stratton St. Margaret, Swindon, Wiltshire. Tel: 0793 822451

j Firth Brown Tools Ltd., Carlisle Street East, P.O. Box 59, Sheffield S4 7QP. Tel: 0742 78500

a,e High Duty Alloys Ltd., 89 Buckingham Avenue Trading Estate, Slough SL1 4PA. Tel: 0753 23901

b,e,g Imperial Metal Industries (Kynoch) Ltd., Birmingham B6 7BA. Tel: 021 356 4848

h,j Kennametal Ltd., 90 Edgbaston Road, Smethwick, Warley, Worcestershire B66 4LB. Tel: 0384 278241

d Mining & Chemical Products Ltd., Alperton, Wembley, Middlesex. Tel: 01 902 1191

d,g,h Murex Ltd., Rainham, Essex. Tel: 04027 53322

a R.T.Z. Aluminium Ltd., Cleveland House, St. James's Square, London S.W.1. Tel: 01 930 7355

e,j Sandvik (UK) Ltd., Manor Way, Halesowen, Birmingham. Tel: 021 550 4700

c,e Titanium International Ltd., Bridge Estate, Thornhill Road, Solihull. Tel: 021 704 3321

h Tungsten Manufacturing Co. Ltd., Fishergate Works, Nr. Brighton, Sussex. Tel: 0273 412281

c Henry Wiggin & Co. ltd., Holmer Road, Hereford. Tel: 0432 6461

b Yorkshire Imperial Metals Ltd., P.O. Box 166, Leeds LS1 1RD. Tel: 0532 701107

CERAMICS

Key to types marketed

a = Alumina
b = Beryllia
c = Borides
d = Boron nitride
e = Porcelain
f = Silicon carbide
  f1 = Hot-pressed
  f2 = Recrystallised
g = Silicon nitride
  g1 = Hot-pressed
  g2 = Reaction-bonded
h = Titania
j = Zirconia

---

g2 Associated Engineering Developments Ltd., Cawston House, Cawston, Rugby, Warwickshire. Tel: 0788 812555

a,b,g2,j Bush Beach Engineering Ltd., Stanley Green Trading Estate, Cheadle Hulme, Cheadle SK8 6RN. Tel: 061 485 8151

a,j Carborundum Co. Ltd., Trafford Park, Manchester M17 1HP. Tel: 061 872 2381

e Doulton Insulators Ltd., Two Gates, Tamworth, Staffordshire. Tel: 0827 62113

a English Glass Co. Ltd., Scudamore Road, Leicester LE3 1OG. Tel: 0533 871371

g Joseph Lucas, Ltd. (Research Centre), Great King Street, Birmingham B19 2XF. Tel: 021 554 5252

a Morgan Matrock Ltd., Stourport-on-Severn, Worcestershire. Tel: 02993 2271

a,e Park Royal Porcelain Co. Ltd., Cox Hill, Sandy, Bedfordshire. Tel: 0767 80305

a,e Royal Worcester Industrial Ceramics Ltd., Tonyrefail, Mid-Glamorgan, South Wales. Tel: 0443 670666

a,e Thermal Syndicate Ltd., P.O. Box 6, Wallsend, Northumberland NE28 6DG. Tel: 0632 625311.

a,e Geo. Wade & Son Ltd., Burslem, Stoke-on-Trent ST6 4AE. Tel: 0782 89321

For advice:

British Industrial Ceramic Manufacturers Federation, Federation House, Station Road, Stoke-on-Trent ST4 2SA. Tel: 0782 48631

British Ceramic Research Association, Queens Road, Penkhull, Stoke-on-Trent ST4 7LQ. Tel: 0782 45431

Institute of Ceramics, Federation House, Station Road, Stoke-on-Trent ST4 2RY. Tel: 0782 44840

Refractories Association of Great Britain, Weston House, West Bar Green, Sheffield S1 2DA. Tel: 0742 48631

COATINGS/ HARDFACINGS

Key to types marketed

a = Plating/solution
b = Paints and lacquers
c = Flame spray coats

---

a,b W. Canning & Co. Ltd., Great Hampton Street, Birmingham B18 6AS. Tel: 021 236 8621

c Cutting and Wear Ltd., Northfield Industrial Estate, Greasebrough Road, Rotherham S60 1QG. Tel: 0709 61041

c Deloro Stellite (BOC Group), Stratton St. Margaret, Swindon. Tel: 0793 822451

a Engelhard Industries Ltd., St. Nicholas Road, Sutton, Surrey. Tel: 01 643 8080

c Greville Hardfacing & Eng. Co. Ltd., Alders Drive, East Moons Moat, Redditch, Worcs B98 0RF. Tel: 0527 25395

a Hilton & Truck Ltd., Roundthorne Estate, Wythenshaw, Manchester M23 9NH. Tel: 061 998 5432

a Ionic Plating Co. Ltd., P.O. Box 59, Grove Street, Smethwick, Warley, West Midlands B66 2QN. Tel: 021 558 2951

a Johnson Matthey Metals Ltd., 100 High Street, London N14 6ET. Tel: 01 882 6111

c Metco Ltd., Chobham, Woking, Surrey. Tel: 09905 7121

c Penistone Hardmetals Co. Ltd., Roman Ridge Road, Sheffield S9 1FH. Tel: 0742 388471

c Technical Hardfacings Ltd., Bank Mill, Manchester Road, Mossley, Lancs. Tel: 045 75 2854

c Union Carbide UK Ltd., Coating Service, Drakes Way, Greenbridge Estate, Swindon, Wilts SN3 3HX. Tel: 0793 29241

c Wall Colmonoy Ltd., Pontardawe, Swansea, West Glamorgan SA8 4HL. Tel: 0792 862287

c Wear Resistance Ltd., Low Road, Oughtibridge, Sheffield S30 3HD. Tel: 074286 2991

ELASTOMERS/RUBBERS AND PLASTICS

B.P. Chemicals International Ltd., Belgrave House, 76 Buckingham Palace Road, London S.W.1. Tel: 01 581 1388

Dow Corning International Ltd., Reading, Berkshire RG1 8PW. Tel: 0734 57251

Du Pont Co. (UK) Ltd., Du Pont House, Fetter Lane, London EC4 1HT. Tel: 01 242 9044

Esso Petroleum Co. Ltd., Victoria Street, London SW1E 5JW. Tel: 01 834 6677

G.E. Co. of New York U.S.A., Birchwood Park, Risley, Warrington WA3 6DA. Tel: 0925 811522

Unitex Ltd., Knaresborough, Yorkshire HG5 0PP. Tel: 0423 862677

Bayer UK Ltd., Bayer House, Richmond TW9 1SJ. Tel: 01 940 6077

Goodyear Tyre & Rubber Co (GB) Ltd., Stafford Road, Bushbury, Wolverhampton WV10 6DH. Tel: 0902 22321

Polysar (UK) Ltd., Imperial Life House, London Road, Guildford, Surrey. Tel: 0483 32551

Lewis & Peat (Rubber) Ltd., 32 St. Maryat Hill, London EC3P 3AJ. Tel: 01 623 9333

Phillips Petroleum Chemicals UK Ltd., Cory House, High Street, Bracknell, Berkshire. Tel: 0344 51711

3M United Kingdom plc, 3M House, P.O. Box 1, Bracknell, Berkshire RG12 1JU. Tel: 0344 26726

Montedison (UK) Ltd., 7–8 Lygon Place, Ebury Street, London SW1. Tel: 01 730 3405

ICI Petroleum & Plastics Division, P.O. Box 6, Bessemer Road, Welwyn Garden City, Hertfordshire AL7 1HD. Tel: 07073 23400

Norsk Hydro Polymers Ltd., Aycliffe Industrial Estate, Darlington, Co. Durham DL5 6EA. Tel: 0325 315122

Norsk Hydro Polymers Ltd., North Street, Havant, Hampshire PO9 1QM. Tel: 0705 486350

Hoechst UK Ltd., Hoechst House, Salisbury Road, Hounslow, Middlesex. Tel: 01 570 7712

BASF United Kingdom Ltd., P.O. Box 4, Earl Road, Cheadle Hulme, Cheadle, Cheshire. Tel: 061 485 6222

Akzo Chemic UK Ltd., Kestrel House, 1–5 Queens Road, Hersham, Surrey. Tel: 09322 47891

Shell Chemicals UK Ltd., 1 Northumberland Avenue, Trafalgar Square, London WC2N 5LA. Tel: 01 438 3000

Ciba-Geigy Plastics & Additives Co., 30 Buckingham Gate, London SW1E 6LH. Tel: 01 828 5676

Borden UK Ltd., North Beddesley, Southampton SO5 9ZB. Tel: 0703 732131

Bakelite UK Ltd., Redfern Road, Tyseley, Birmingham B11 2BJ. Tel: 021 706 3322

TBA Industrial Products, P.O. Box 40, Rochdale, Lancashire OL12 7EQ. Tel: 0706 47422

Industrial Polymers (U.K.) Ltd., Church Hill, Orpington, Kent BR6 0HE. Tel: 0689 36224

Borg-Warner Chemicals, Bo'Ness Road, Grangemouth FK3 9XF. Tel: 03244 483490

Monsanto Ltd., 14 Tothill Street, London SW1H 9LH. Tel: 01 222 5678

DSM United Kingdom Ltd., Kingfisher House, Kingfisher Walk, Redditch, Worcestershire B97 4EZ. Tel: 0527 68254

Scott Bader Co. Ltd., Wollaston, Wellingborough, Northamptonshire NN9 7RL. Tel: 0933 663100

Freeman Chemicals Ltd., P.O. Box 8, Ellesmere Port, South Wirral L65 4AH. Tel: 051 355 6171

For advice:

Rubber and Plastics Research Association of Great Britain, Shawbury, Shrewsbury, Salop SY4 4NR. Tel: 0939 250383

Bibliography

Bisson, E. E. (1969) "Various modes of wear and their controlling factors", *Evaluation of Wear Testing*, A.S.T.M. STP 446, pp. 1–22.

Bowden, F. P. and Tabor, D. (1964) *The Friction and Lubrication of Solids*, Clarendon Press, Oxford.

Buckley, D. H. (1972) *A Fundamental Review of the Friction and Wear Behaviour of Ceramics*, N72–23582 (NASA-TM-X-68046).

Colbalt Monograph (1960) Centre d'information du Cobalt, Brussels.

Colombier, L. and Hochmann, J. (1967) *Stainless and Heat Resisting Steels*, Edward Arnold (Publishers) Ltd.

Conte, A. A., Jnr. (1973) "Coatings to improve wear resistance", *Mechanical Engineering*, Jan. 1973, pp. 18–24.

Dawihl, W., and Frisch, B. (1960) "Wear properties of tungsten carbide and aluminium oxide sintered materials", *Wear*, **12**, 17–25.

Eyre, T. S., Iles, R. F., and Gasson, D. W. (1969) "Wear characteristics of flake and nodular graphite cast iron", *Wear*, **13**, 229–245.

Fairhurst, W. *Chromium-Molybdenum White Cast Irons for Abrasive Applications*. Climax Molybdenum Co. Ltd.

Finnie, I. (1962) "Erosion by solid particles in a fluid stream", ASTM, Special Technical Publication, No. 307. Symposium on erosion and cavitation, pp. 70–82.

Hirst, W. (1965) "Wear", *Metallurgical Reviews* **10**, 145–172.

Hurricks, P. L. (1972) "The fretting wear of mild steel from room temperature to 200°C", *Wear*, **19**, 207–229.

Hurricks, P. L. (1973) "Some metallurgical factors controlling the adhesive and abrasive wear resistance of steels—a review", *Wear*, **26**, 285–304.

Hurricks, P. L. (1971), "Overcoming industrial wear", *Industrial Lubrication and Tribology*, Oct. 1971, pp. 345–356.

Hurricks, P. L. (1972) Review paper: "Some aspects of the metallurgy and wear resistance of surface coatings", *Wear*, **22**, 291–320.

Lancaster, J. K. (1968/69) "Relationships between the wear of polymers and their mechanical properties", *Proc. Inst. Mech. Engrs.*, Paper 12, Vol. 183, Pt. 3P, pp. 98–106.

Lansdown, A. R. (1982) *Lubrication: a Practical Guide to Lubricant Selection*, Pergamon Press, Oxford.

Lipson, C. (1967) "Wear considerations in design", *Studies in Design Engineering*, Prentice-Hall Inc., New Jersey.

Tribology Handbook (1973) Ed. M. J. Neale. Butterworths, London.

Rabinowicz, E. (1965) *Friction and Wear of Materials*, John Wiley & Sons, New York.

Richardson, R. C. D. (1967) "The wear of metals by hard abrasives", *Wear*, **10**, 291–309.

Rollason, E. C. (1973) *Metallurgy for Engineers* (4th ed.), Edward Arnold (Publishers) Ltd.

Sibley, L. B. and Allen, C. M. (1962) "Friction and wear behaviour of refractory materials at high sliding velocities and temperatures", *Wear*, **5**, 312–329.

Suh, N. P. (1973) "The delamination theory of wear", *Wear*, **25**, 111–124.

Wahl, W. (1973) "Wear-resistant metallic cast materials", *V.D.I., Berichte*, **194**, 79–94, Risley Trans. ex. N.L.L.

Wear Control Handbook (1980) Ed. Peterson, M. B. and Winer, W. O. ASME, New York.

Index